AKADEMIE DER WISSENSCHAFTEN DER DDR

Zentralinstitut für Physik der Erde (ZIPE)

Seismological Bulletin 1974
Station Moxa (MOX)

By

Johannes Stelzner, Dorothea Güth,
and Joachim Weyrauch

With 1 Figure

AKADEMIE-VERLAG · BERLIN
1979

Erschienen im Akademie-Verlag, DDR-108 Berlin, Leipziger Straße 3—4

Lizenznummer: 202 · 100/465/79
Gesamtherstellung: Druckerei „Thomas Müntzer", 582 Bad Langensalza
Bestellnummer: 762 731 0 (2004/B/1974) · LSV 1437
Printed in GDR
DDR 36,— M

PREFACE

The annual Seismological Bulletin 1974 for the Station Moxa (MOX) continues the series of publications about seismological records at the Station Moxa edited by the Central Earth Physics Institute of the Academy of Sciences of the German Democratic Republic.

The provisional analysis of the records of station Moxa was performed in the Seismological Service under the direction of JOHANNES STELZNER by JOACHIM WEYRAUCH and BRIGITTE HÄNSCH.

The annual Bulletin 1974 was prepared by DOROTHEA GÜTH and JOACHIM WEYRAUCH with the technical assistance of URSULA DÖRING.

Control of the instruments of the station Moxa was carried out under the direction of CHRISTIAN TEUPSER.

H. KAUTZLEBEN
Director

TABLE OF CONTENTS

Preliminary notes for the interpretation of seismograms

In the Bulletin the international code is used:

1. Phase interpretation

Pg	— direct longitudinal wave in near epicentral distances (D < 10°)
Pb, Pn	— guided longitudinal head waves along the CONRAD- or MOHOROVIČIĆ-discontinuity (D < 10°)
P	— direct longitudinal wave travelled through the earth mantle
P diff	— direct longitudinal wave diffracted around the core boun dary
PKIKP	— direct longitudinal wave travelled through the inner core (travel-time branch DF)
PKHKP	— direct longitudinal wave refracted in the intermediary zone between inner and outer core. Phase symbol according to BOLT [1] (travel-time branch GH)
PKP2	— direct longitudinal wave travelled through the outer core only (travel-time branch AB)
PKP	— first noticeable onset of longitudinal core phase not identified
PP, PPP	— waves reflected at the earth surface with permanent longitudinal character
PKKP	— core phase reflected once within the core at the outer core boundary
PKPPKP	— longitudinal core phase reflected at the earth surface
Sg	— direct transversal wave in near epicentral distances (D < 10°)
Sb, Sn	— guided transversal head waves along the CONRAD- or MOHOROVIČIĆ-discontinuity (D < 10°)
S	— direct transversal wave travelled through the earth mantle

SKS — direct wave travelled transversal through the mantle and longitudinal through the core

SS, SSS — waves reflected at the earth surface with permanent transversal character

SKKS — wave travelled transversal through the mantle, longitudinal through the core and reflected within the core at the outer core boundary

PcP, ScS, PcS, ScP — longitudinal and transversal waves with steady or changing character reflected at the outer core

PS, SP, PPS — longitudinal and transversal waves with changing character reflected at the surface of the earth

pP, sP. pPP, sPP, pPKIKP, sPKP2, pS — phases of deep-focus earthquakes of longitudinal or transversal waves with steady or changing character. p;s — reflected near the epicentre

pPKP, sPKP — phases of deep focus earthquakes of longitudinal core waves not exactly to be coordinated

SKP, PKS — core phases with different character before and after the direct transit of the core

SKSP — SKS wave with longitudinal character after the reflection at the surface of the earth

P1, P2, P3, . . ., S1, S2, . . . — multiple onsets of body waves

Pn, Sn — teleseismic Pn and Sn waves in the epicentral distances $23° < D < 40°$ after BÅTH [2]

Pa, Sa — waves probably guided in the astenosphere channel or higher modes of surface waves

PL — leaking modes, normal dispersed train of waves of periods greater than about 10 s, beginning at or near the time of initial P-wave

X, Y, Z — remarkable phases of body waves, not to be identified

LmV, LmH — maximum of the vertical and horizontal component respectively of longperiodical surface waves. If there are several maxima with comparable proportions in A/T, the numeration was carried out in a temporal sequence e.g. Lm1H, Lm2H

The phase symbol is followed by the designation of the type of seismometer from which the time of onsets is taken.

A — seismograph with amplitude characteristic of type A (short-period)

B — seismograph with amplitude characteristic of type B (middle-period)

C — seismograph with amplitude characteristic of type C (long-period)

2. *Measurements of amplitudes and calculation of magnitudes*

All data of amplitudes and periods printed in the column "remarks" are always taken from the records of the same instruments, from which are taken the onset-times of the corresponding phases. The symbol of phase and component is followed by the symbol of the type of instruments e.g.: PV A, PV B, LmH B. LmV C.

Data of amplitudes obtained from records of instruments of type A are given in units of length of nm (1 nm = 1 nanometre = 10^{-6} millimetre). Data of amplitudes obtained from instruments of type B and such obtained from instruments of type C are given in units of length μm (1 μm = 1 micrometre = 10^{-3} millimetre) e. g.: PVA 1.3 s 38.6 nm, SHB 10 s 3.2 μm, LmH B 22 s 15 μm.

Magnitudes are determined from all those phases, for which calibrating functions are known and internationally used, i. e.

for maxima of body waves P(PH, PV), PP(PPH, PPV), and S(SH)-Q-functions from Gutenberg and Richter [3] — and

for maxima of surface waves (h < 100 km) LmH, LmV — calibrating functions from Prague σ [4] —.

The station correction S was not yet taken into consideration.

MB — magnitude of vertical component V of the first onset of P-waves given by Neis

MS — magnitude of horizontal component H of the maximum surface wave given by Neis

M — magnitude calculated from given data of station Moxa. Notice the wave type and the type of instruments written on the same line

3. *Direction of body-wave onsets*

If the direction of motion at the beginning of a wave onset is clearly to be recognized, the sign + or — is placed before the phase symbol. It means:

in the Z component	+	ground motion upwards, compression
	—	ground motion downwards, dilatation
in the N component	+	ground motion to the north
	—	ground motion to the south
in the E component	+	ground motion to the east
	—	ground motion to the west

4. *Further abbreviations*

i — sharp beginning of phase motion (impetus)

e — gradual beginning of phase motion (emersio)

D — epicentral distances in degree (°), calculated according to geocentric coordinates, the maximum error of the own calculations amounts to ± 0,1°.

Az — azimuth: clockwise measured angle between north direction in epicentre and the connecting line from epicentre to station Moxa

h — depth of focus in km, our data for depth of focus are based on travel-time curves for deep focus earthquakes after GUTENBERG and RICHTER [5]

H — origin time in UTC (Universal Time)

NEIS — National Earthquake Information Service, Denver, Colorado, USA

BCIS — Bureau Central International de Seismologie, Strasbourg, France,

ANUSSR — Akademia Nauk USSR, Moscow, USSR

AEC — United States Atomic Energy Commission, Washington, D. .

ISC — International Seismological Centre, Newbury, UK

NORSAR — Norwegian Seismic Array, Kjeller, Norway

For abbreviations of seismological stations and other agencies in the international three letter code see the introductions to the Regional Catalogue of Earthquakes, Edinburgh and the Bulletins of the International Seismological Centre, Edinburgh.

Round brackets indicate uncertainties in interpretation of phase, time, depth of focus or epicentral distances, respectively.

[1] BOLT, A., The velocity of seismic waves near the earths center. Bull. Seism. Soc. Am. 54 (1964) 1, 191—208.

[2] BÅTH, M., Propagatíon of Sn and Pn teleseismic distances. Pure and Applied Geophysics 65 (1966/II) 19—30.

[3] GUTENBERG, B. and RICHTER, C. F., Magnitude and energy of earthquakes. Annali di Geofisica 9 (1956) 1, 1—15.

[4] KÁRNÍK, V., KONDORSKAJA, N. V. u. a., Standardization of the earthquake magnitude scale. Stud. Geophys. et Geodet., Prague 6 (1962) 41—48.

[5] GUTENBERG, B. and RICHTER, C. F., Materials for the study of deep-focus earthquakes. Bull. Seism. Soc. Am. 26 (1936) 4, 341—390.

Seismological Station Moxa (MOX)
of the Central Earth Physics Institute

Elevation above mean sea level:	**455 m**
Bedrock:	clay slate of the lower carboniferous formation
Geographic coordinates:	**$\varphi = 50°38'46''$N $\lambda = 11°36'58''$E**
Address:	**Central Earth Physics Institute Seismological Service DDR-69, Jena, Burgweg 11 German Democratic Republic Telex: 05886275 seis dd**

Seismographs and their parameters 1974

T_s — seismometer free period
T_g — galvanometer free period
D_s — seismograph damping
D_g — galvanometer damping
V_0 — magnification factor
N — north-south component
E — east-west component
Z — vertical component
σ^2 — coupling coefficient

SKM — Seismograph Kirnos modified
SSJ — Seismic Station Apparatus Type Jena
VSJ — Vertical Seismograph Type Jena

	Type of Seismograph	Comp.	T_s[s]	T_g[s]	D_s	D_g	V_0	σ^2
	VSJ II	Z	0.23	0.065	0.33	1.2	300000	0.048
	VSJ II	Z	1.0	1.0	0.5	0.5	47200	0.56
A	SKM III	N	1.64	0.39	0.52	1.97	24000	0.047
		E	1.63	0.40	0.50	1.92	24700	0.047
		Z	1.64	0.39	0.51	1.96	23400	0.050
B	SSJ I	N	20	1.13	0.50	8.87	109	0.077
		N	20	1.14	0.50	8.79	1110	0.077
		E	20	1.13	0.49	8.85	103	0.070
		E	20	1.16	0.49	8.61	1050	0.070
		Z	20	1.13	0.48	8.82	108	0.057
		Z	20	1.24	0.48	8.05	910	0.057
C	SSJ I/L	N	30	89.5	1.49	0.5	1000	0.085
	(until March. 12.)	E	30	74.4	1.24	0.5	1010	0.051
		Z	30	91.0	1.52	0.5	1060	0.108
C	SSJ I/L	N	30	87.5	1.46	0.5	1050	0.103
	(from July. 18.)	E	30	75.8	1.26	0.5	1070	0.056
		Z	30	87.7	1.46	0.5	1040	0.094
	STRAIN/L	N		85.4		0.70	65*	
		E		86.2		0.70	67*	
	(coupled)	N + E		86.2		0.70	42*	

*) for apparent wave velocity 5 km s^{-1}

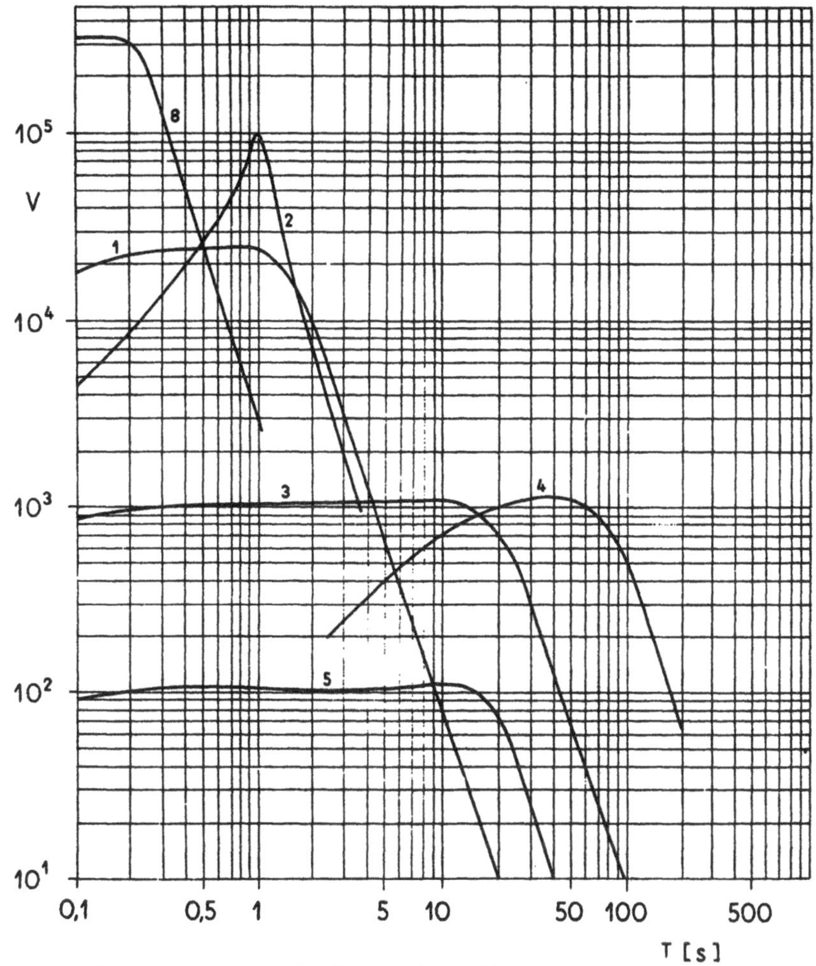

Fig. 1. Station Moxa, mean amplitude characteristics 1974

1 — Seismograph Kirnos Modified-III (SKM-III) (NS-, EW- and Z-component)

2 — Seismograph Type Jena II (Z-component)

3 — Seismic Station Apparatus Type Jena I/1000 (SSJ-I/1000) (NS-, EW- and Z-component)

4 — Seismic Station Apparatus Type Jena I/L (SSJ-I/L) (NS-, EW- and Z-component)

5 — Seismic Station Apparatus Type Jena I/100) (SSJ-I/100) (NS-, EW- and Z-component)

6 — Seismograph Type Jena II (Z-component)

Seismological Recordings at Station Moxa 1974

Day	Phase		h m s	Remarks
1.	ePKHKP	A	06 07 14	Tonga Islands 20.07 S 174.25 W H = 05 47 27.6 h = normal MB = 4.6 D = 149.14 Az = 352.7 (NEIS)
1.	eP ePP	A A	08 09 56.5 13 54.5	Mariana Islands Region 21.63 N 142.85 E H = 07 57 04.5 h = 333 km MB = 5.0 D = 96.19 Az = 331.2 (NEIS) PPV A 1.7s 48.5nm M = 5.3
1.	ePKHKP	A	09 48 12.5	Fiji Islands Region 21.93 S 176.93 W H = 09 28 43.1 h = 194 km MB = 4.8 D = 150.57 Az = 348.9 (NEIS)
1.	ePKIKP iPKHKP iPKP2 pPKP pPKP2	A A A A A	13 02 06 02 13.5 02 25 04 03 04 20.5	South of Fiji Islands 23.70 S 179.92 E H = 12 43 16.1 h = 501 km MB = 5.0 D = 151.60 Az = 344.3 (NEIS) PKHKPV A 1.3s 39.3nm PKP2V A 1.4 60.5nm
1.	eP	A	14 03 57	Burma-India Border Region 23.9 N 94.1 E H = 13 51 33 h = 285 km MB = 4.1 D = 67.24 Az = 317 (ISC)
1.	eP e	A A	14 19 59.5 20 25	Northern Sumatra 4.63 N 95.90 E H = 14 07 40.1 h = 59 km MB = 5.4 D = 82.81 Az = 320.3 (NEIS) PV A 1.4s 18.6nm M = 4.9
2.	ePKIKP	A	00 21 03.5	New Hebrides Islands 14.60 S 167.35 E H = 00 01 53.5 h = 176 km MB = 5.0 D = 139.03 Az = 336.5 (NEIS) PKIKPV A 1.4s 18.6nm
2.	ePKP	A	05 14 18	Samoa Region 15.15 S 172.1 W H = 04 54 45.3 h = 33 km MB = 4.9 D = 144.46 Az = 356 (ISC)

Day	Phase		h m s	Remarks
2.	eP	ABC	10 56 09	Northern Chile 22.54 S 68.40 W
	e(sP)	ABC	56 42	H = 10 42 29.9 h = 105 km MB = 6.4
	ePP	ABC	11 00 17	D = 101.06 Az = 39.7 (NEIS)
	eSKS	BC	06 42	PV A 1.6s 110.0nm M = 6.2
	eS	B	07 40	LmH B 20 54.4 µm
	eSP	B	09 10	LmV B 20 50.5 µm
	ePPS	B	10 16	
	iPKKP	AB	12 48	
	eSS	B	14 46	
	ePKPPKP	A	20 40	
	LmH	B	39.6	
	LmV	B	39.8	
2.	eP	ABC	14 53 54	Northeast of Taiwan 26.02 N 124.38 E
	epP	A	54 45	H = 14 41 47.9 h = 205 km MB = 5.7
	esP	B	5506	D = 83.41 Az = 323.8 (NEIS)
				h = 212 km
				PV A 1.8s 291.0nm M = 5.7
2.	e(P)	A	19 32 08	Central Italy 43.23 N 12.9 E
				H = 19 29 39 h = 0 km (ISC)
				D = 9.3
3.	eP	A	04 06 11.5	Mariana Islands 20.54 N 145.05 E
				H = 03 52 50.4 h = 139 km MB = 5.1
				D = 98.12 Az = 332.2 (NEIS)
3.	eP	A	07 43 28	Turkey 39.74 N 26.82 E
	LmH	B	48.6	H = 07 39 48 h = 29 km MB = 4.2
	LmV	B	50.0	D = 15.25 Az = 321 (ISC)
				PV A 1.1s 28.2nm M = 4.4
3.	ePKP	A	11 34 54.5	New Hebrides Islands 14.56 S 166.35 E
				H = 11 15 24.0 h = 9 km MB = 5.5
				D = 138.60 Az = 335.7 (NEIS)
				PKPV A 1.2s 20.3nm

Day	Phase		h m s	Remarks
3.	e	A	22 28 07.5	
	e	A	29 25	
	e	A	29 53	
4.	eP	A	09 36 17	Kirgiz-Sinkiang Border Region
	e	A	36 26	40.61 N 77.65 E
	LmH	B	55.0	H = 09 27 55.7 h = 26 km MB=5.5 MS=4.8
	LmV	B	56.7	D = 45.82 Az = 305.8 (NEIS)
				PV A 1.6s 38.5nm M = 5.1
4.	ePKIKP	A	19 01 25.5	New Hebrides Islands 14.84 S 166.53 E
				H = 18 41 56.3 h = 21 km MB = 5.3
				D = 138.93 Az = 335.7 (NEIS)
5.	eP	A	06 48 25	Central Italy 43.20 N 12.76 E
				H = 06 45 15.8 h = 0 km (ISC)
				D = 12.45
5.	eP	A	07 34 05.5	South Atlantic Ridge 32.4 S 13.17 W
				H = 07 21 29.3
				D = 85.56 Az = 16 (ISC)
5.	eP	A	07 35 22	Central Italy 43.43 N 12.30 E
				H = 07 33 37.3 h = normal MB = 3.9
				D = 7.23 Az = 356.5 (NEIS)
5.	iPKP	A	08 22 37	New Hebrides Islands 20.03 S 170.14 E
				H = 08 03 01.3 h = normal MB = 5.0
				D = 145.04 Az = 336.0 (NEIS)
				PKPV A 1.3s 32.8nm
5.	eP	A	08 47 17.5	Near Coast of Peru 12.30 S 76.35 W
	epP	A	47 44	H = 08 33 50.7 h = 98 km MB = 6.3
	eS	BC	58 40	D = 98.11 Az = 40.0 (NEIS)
	LmH	B	09 15.7	PV A 1.6s 74.2nm M = 6.0
	LmV	B	36.4	LmH B 21 4.2 µm
				LmV B 16 2.2 µm

Day	Phase		h m s	Remarks
5.	eP	A	14 12 50.5	Fox Islands, Aleutian Is. 52.17 N 171.42 W H = 14 00 56.8 h = 41 km MB=5.4 MS=4.7 D = 77.53 Az = 358.0 (NEIS) PV A 0.9s 42.8nm M = 5.5
5.	eP	A	15 49 44	Off Coast of Oregon 42.63 N 126.42 W H = 15 37 33.7 h = normal MB=4.6 MS=4.7 D = 80.15 Az = 25.6 (NEIS)
5.	eP	A	16 06 13.5	Off Coast of Oregon 42.48 N 126.60 W
	ePcP	A	06 21.5	H = 15 54 03.3 h = normal MB=4.9 MS=5.1 D = 80.35 Az = 25.5 (NEIS) PV A 1.8s 60.8nm M = 5.3
5.	e(PKIKP)	A	22 02 55	New Hebrides Islands 14.66 S 166.64 E H = 21 43 19.1 h = 18 km MB = 5.1 D = 138.81 Az = 335.9 (NEIS)
5.	e	A	23 25 30	
5.	eP	A	23 41 29	Off Coast of Oregon 42.59 N 126.58 W
	e	A	41 36	H = 23 29 18.6 h = 22 km MB=5.0 MS=4.6 D = 80.25 Az = 25.5 (NEIS) PV A 1.6s 38.5nm M = 5.2
6.	ePKIKP	A	01 48 28	Tonga Islands 18.39 S 173.77 W H = 01 28 53.4 h = normal MB = 4.7 D = 147.53 Az = 353.6 (NEIS)
6.	ePKHKP	A	04 39 44	Tonga Islands 21.85 S 175.06 W H = 04 19 52.7 h = normal MB=5.0 MS=5.3 D = 150.78 Az = 351.3 (NEIS) PKHKPV A 2.0s 85.5nm
6.	eP	A	10 12 52	North Atlantic Ocean 57.52 N 33.77 W
	LmV	B	23.9	H = 10 07 12.6 h = normal MB=4.9 MS=4.7
	LmH	B	24.2	D = 27.06 Az = 85.0 (NEIS) PV A 2.0s 77.0nm M = 5.0 LmV B 14 2.8/um 5.1

Day	Phase		h m s	Remarks
6.	eP	A	10 38 45.5	North Atlantic Ocean 57.73 N 33.57 W H = 10 33 06.1 h = normal MB=4.6 MS=4.3 D = 26.94 Az = 85.6 (NEIS) PV A 1.3s 26.2nm M = 4.7
6.	eP1 eP2	A A	14 42 22 42 26	North of Ascension Island 1.50 S 15.53 W H = 14 32 39.9 h = normal MB=5.3 MS=4.9 D = 56.89 Az = 20.3 (NEIS) P1V A 2.0s 51.3nm M = 5.2 P2V A 1.5 60.3nm 5.4
6.	ePKIKP	A	15 33 27.5	Kermadec Islands 30.38 S 177.80 W H = 15 13 31.9 h = 19 km D = 158.56 Az = 343.4 (NEIS)
6.	ePKIKP	A	17 58 50	New Hebrides Islands 14.86 S 167.24 E H = 17 39 44.1 h = 117 km MB = 5.6 D = 139.22 Az = 336.3 (NEIS) PKIKPV A 1.2s 12.2nm
7.	eP	A	04 08 22.5	Philippine Islands Region 19.13 N 121.08 E H = 03 55 39.4 h = 39 km MB=5.0 MS=5.2 D = 87.10 Az = 323.1 (NEIS)
7.	esP	A	05 29 53.5	Burma-India Border Region 23.52 N 94.81 E H = 05 18 24.3 h = 105 km MB = 4.9 (NEIS) D = 68.3
7.	eP ePcP	A A	08 37 58 38 22	Southern Alaska 59.81 N 153.72 W H = 08 27 03.5 h = 128 km MB = 4.9 D = 69.26 Az = 9.9 (NEIS) PV A 0.8s 13.5nm M = 4.8
7.	eP e	A A	15 31 03.5 31 11.5	Western Iran 33.30 N 47.92 E H = 15 24 38.2 h = 32 km MB = 5.0

Day	Phase		h m s	Remarks
cont. 7.	LmH	C	15 43.3	D = 31.67 Az = 314.1 (NEIS)
	LmV	C	46.3	LmH C 24s 1.5 µm M = 4.6
				LmV C 18 1.3 µm 4.8
7.	eP	A	16 49 54	Tucuman Province, Argentina
	e	A	54 04	26.89 S 65.70 W
	LmH	B	17 36.7	H = 16 35 57.8 h = 33 km MB = 5.8
	LmV	B	36.8	D = 102.83 Az = 39.6 (NEIS)
				LmH B 18s 1.6 µm M = 5.6
				LmV B 18 1.6 µm 5.6
7.	eSg	A	16 55 42	Austria 47.5 N 14.9 E
				Explosion 9.75 t
				H = 16 53.7 (VIE)
				D = 3.82 Az = 327 (ISC)
8.	eP1	A	22 00 39	Atlantic-Indian Rise 38.95 S 46.18 E
	eP2	A	00 41	H = 21 47 21.7 h = normal MB=6.0 MS=6.1
	eSKS	C	11 16	D = 94.25 Az = 338.8 (NEIS)
	eS	BC	11 54	P1V A 2.0s 42.8nm M = 5.5
	ePS	BC	13 12	P2V A 3.0 224.0nm 6.1
	eSS	BC	18 20	LmH B 16.5 6.3 µm 6.2
	LmV	B	49.8	LmV B 16.5 6.6 µm 6.2
	LmH	B	50.0	
9.	eP	A	03 01 23	Off East Coast of Kamchatka
	LmV	B	40.2	51.65 N 159.64 E
	LmH	B	40.3	H = 02 49 46.3 h = normal MB=5.4 MS=5.4
				D = 74.54 Az = 339.5 (NEIS)
				PV A 1.3s 56.8nm M = 5.4
				LmH B 15 1.9 µm 5.5
				LmV B 15 1.9 µm 5.6
9.	eP	A	13 12 23	Albania 41.99 N 19.02 E
	eS	A	14 10	H = 13 09 55.8 h = 54 km
				D = 10.05 Az = 332 (ISC)

Day	Phase		h m s	Remarks
9.	eP	A	16 08 34	Eastern Mediterranean Sea
				34.6 N 30.8 E
				H = 16 03 45 h = 0 km MB = 4.1
				D = 21.24 Az = 325 (ISC)
				PV A 1.0s 11.8nm M = 4.2
9.	eP	A	21 18 10.5	Iran 30.57 N 57.57 E
				H = 21 10 41.4 h = normal
				D = 39.47 Az = 314.0 (NEIS)
10.	ePKHKP	A	00 58 45.5	South of Fiji Islands 22.03 S 179.71 W
	ePKP2	A	58 53.5	H = 00 39 56.0 h = 546 km MB = 4.6
				D = 150.09 Az = 345 (ISC)
				PKHKPV A 0.9s 19.5nm
10.	eP	A	02 48 37.5	Off East Coast of Kamchatka
	LmV	B	03 27.5	51.65 N 159.45 E
	LmH	B	27.6	H = 02 37 03.2 h = 41 km MB = 4.9
				D = 74.49 Az = 339 (ISC)
				LmH B 16s 1.1 μm M = 5.3
				LmV B 16 1.4 μm 5.4
10.	eP	A	05 30 31	Off East Coast of Kamchatka
	LmV	B	06 09.4	51.54 N 159.72 E
	LmH	B	09.5	H = 05 18 53 h = 24 km MB = 5.2
				D = 74.66 Az = 340 (ISC)
				PV A 1.5s 30.2nm M = 5.1
				LmH B 14 1.6 μm 5.5
				LmV B 15 2.0 μm 5.6
10.	ePKHKP	A	09 10 29.5	New Hebrides Islands 14.43 S 166.86 E
	ePKIKP	A	10 33.5	H = 08 51 13.3 h = 34 km MB=6.7 MS=7.2
	ePP	ABC	13 27	D = 138.69 Az = 336.2 (NEIS)
	ePKKS	B	23 24	PKIKPV A 1.7s 24.4nm
	eSPP	B	25 32	PKHKPV A 2.0 205.0nm
	e(SS)	B	32 12	LmH B 20 70.5 μm M = 7.4
	eSSS	B	37 20	LmV B 18 58.2 μm 7.4
	LmH	B	10 13.7	
	LmV	B	20.1	

Day	Phase		h m s	Remarks
10.	ePKIKP	A	10 20 15	New Hebrides Islands 13.85 S 166.63 E
				H = 10 00 53.8 h = 66 km MB = 5.7
				D = 138.07 Az = 336.3 (NEIS)
				PKIKPV A 1.8s 60.8nm
10.	eP	A	11 56 19	Sicily 38.86 N 14.83 E
	e	A	57 22	H = 11 53 33 h = 337 km MB = 4.0
				D = 12.00 Az = 350 (ISC)
10.	eP	A	12 57 06	Near East Coast of Honshu, Japan
				36.34 N 141.67 E
				H = 12 44 44.8 h = 47 km MB = 5.1
				D = 82.86 Az = 330.6 (NEIS)
10.	epP	A	16 25 57	Kyushu, Japan 31.79 N 131.70 E
	LmH	B	17 06.6	H = 16 13 25.8 h = 37 km MB=5.2 MS=5.1
	LmV	B	06.8	D = 82.39 Az = 326.2 (NEIS)
				pPV A 1.2s 12.2nm
				LmH B 17 1.9 μm M = 5.5
				LmV B 16 2.2 μm 5.7
10.	eP	A	22 37 27	North Atlantic Ocean 57.33 N 33.57 W
	e	A	37 41	H = 22 31 47.8 h = normal MB=5.1 MS=4.6
	LmH	B	48.7	D = 26.97 Az = 84.8 (NEIS)
	LmV	B	48.7	PV A 2.0s 111.0nm M = 5.1
				LmH B 15 1.3 μm 4.6
				LmV B 14 2.0 μm 5.0
11.	eP	A	02 11 45	Hindu Kush Region 36.30 N 70.77 E
				H = 02 03 50.4 h = 159 km MB = 4.8
				D = 44.09 Az = 308.2 (NEIS)
				PV A 1.4s 34.9 M = 4.8
11.	ePKIKP	A	05 55 57.5	New Hebrides Islands 14.16 S 166.59 E
	ePKS	A	59 34.5	H = 05 36 30.8 h = 15 km MB=5.7 MS=6.2
	LmV	B	06 59.1	D = 138.33 Az = 336.1 (NEIS)
	LmH	B	59.6	PKIKPV A 1.0s 27.6nm
				LmH B 22 7.2 μm M = 6.3
				LmV B 20 5.8 μm 6.3

Day	Phase		h m s	Remarks
11.	eP	A	09 34 02.5	Off East Coast of Kamchatka 51.68 N 159.49 E H = 09 22 25.5 h = normal MB = 4.6 D = 74.47 Az = 339.4 (NEIS)
11.	eP	A	15 12 07	South Atlantic Ridge 26.36 S 13.23 W H = 15 00 01.2 h = normal MB = 4.8 D = 79.76 Az = 15.8 (NEIS)
11.	e	A	21 18 32	Aegean Sea 40.14 N 24.46 E H = 21 15 06.1 h = 45 km MB = 3.8 D = 13.83 Az = 323.7 (NEIS)
12.	ePKP	A	06 40 09	Fiji Islands Region 18.45 S 173.38 E H = 06 20 29.2 h = normal MB = 4.9 D = 144.74 Az = 339.8 (NEIS)
12.	ePb	A	14 49 04	Northern Italy 44.15 N 10.19 E H = 14 47 04.4 h = 0 km D = 6.57 Az = 8 (ISC)
12.	eP	A	20 12 43	Panama 9.95 N 78.74 W H = 20 00 21.6 h = 37 km MB=4.8 MS=4.0 D = 82.62 Az = 39.9 (NEIS)
13.	e	A	13 40 16.5	
13.	ePKIKP ePKS	A A	18 12 15 16 09.5	New Hebrides Islands 14.41 S 166.36 E H = 17 52 45.7 h = normal MB=5.0 MS=4.6 D = 138.47 Az = 335.8 (NEIS)
13.	eP	A	21 41 37.5	Central Mid-Atlantic Ridge 3.66 N 31.57 W H = 21 31 39.4 h = normal MB=5.2 MS=4.8 D = 59.19 Az = 30.5 (NEIS) PV A 2.3s 110.0nm M = 5.6

Day	Phase		h m s	Remarks
14.	ePn	A	08 48 31	Northern Italy 44.8 N 8.7 E
	e	A	48 48	H = 08 46 53 h = 0 km
				D = 6.13 Az = 17 (ISC)
14.	eP	A	20 43 31	Kurile Islands 48.76 N 154.96 E
				H = 20 31 43.0 h = 14 km MB=5.5 MS=4.8
				D = 76.10 Az = 336.9 (NEIS)
				PV A 1.6s 52.2nm M = 5.4
14.	e(pP)	A	21 37 08	Off East Coast Kamchatka 51.69 N 159.4 E
				H = 21 25 24 h = 55 km MB = 4.3
				D = 74.44 Az = 339 (ISC)
14.	ePKP	A	23 51 53	Solomon Islands 9.64 S 161.16 E
				H = 23 32 10.6 h = 53 km MB = 5.6 (NEIS)
				D = 132.1
15.	ePKP2	A	08 52 34.5	Kermadec Islands 30.92 S 178.93 W
	e	A	52 40	H = 08 32 14.0 h = 114 km MB = 5.6
				D = 158.78 Az = 341.2 (NEIS)
				PKP2V A 1.4s 55.8nm
15.	e(P)	A	17 37 55.5	Southern Greece 37.41 N 21.01 E
				H = 17 34 19.4 h = 38 km MB = 4.2
				D = 14.84 Az = 336.1 (NEIS)
15.	eP	A	19 52 13	Iceland 64.76 N 17.11 W
	LmH	B	20 01.3	H = 19 47 34.3 h = normal MB = 4.7
	LmV	B	01.4	D = 20.60 Az = 119.6 (NEIS)
				PV A 1.6s 44.0nm
				LmH B 17 1.3/um M = 4.4
				LmV B 16 0.7/um 4.3
15.	ePg	A	20 12 41.5	Switzerland 47.07 N 7.16 E
				H = 20 11 13.3 h = normal
				D = 4.64 Az = 37.8 (NEIS)

Day	Phase		h m s	Remarks
15.	eP	A	23 01 20	<u>Szechwan Province, China</u>
	LmH	B	27.6	32.91 N 104.20 E
	LmV	B	33.4	H = 22 50 29.9 h = normal MB=5.3 MS=5.7
				D = 66.87 Az = 316.2 (NEIS)
				PV A 1.7s 42.4nm M = 5.3
				LmH B 17.5 2.7 μm 5.5
				LmV B 14 1.6 μm 5.4
16.	eP	A	05 04 55	<u>Taiwan</u> 23.46 N 121.43 E
	LmH	B	47.3	H = 04 52 30.7 h = 58 km MB = 5.1
	LmV	B	47.3	D = 83.85 Az = 323.0 (NEIS)
				PV A 1.0s 19.7nm M = 5.1
				LmH B 16 1.5 μm
				LmV B 14.5 1.3 μm
17.	eP	A	02 56 54.5	<u>Luzon, Philippine Islands</u>
				16.80 N 119.96 E
				H = 02 44 09.0 h = 73 km MB = 5.1
				D = 88.31 Az = 323 (ISC)
				PV A 2.0s 42.7nm M = 5.2
17.	eP	A	08 51 00.5	<u>Kurile Islands</u> 43.79 N 147.21 E
				H = 08 39 06.7 h = 77 km MB = 5.2
				D = 78.32 Az = 332.9 (NEIS)
				PV A 1.1s 26.2nm M = 5.0
17.	ePP	A	13 10 39.5	<u>Bali Sea</u> 7.75 S 117.50 E
				H = 12 52 25.5 h = 257 km MB = 5.6
				D = 106.02 Az = 320.4 (NEIS)
17.	e	A	15 39 53	
18.	ePKP2	A	07 13 00.5	<u>West of Macquarie Islands</u>
	LmH	B	08 35.5	60.30 S 150.45 E
	LmV	B	37.3	H = 06 52 43.4 h = normal MB 5.7 MS=5.6
				D = 155.15 Az = 265.7 (NEIS)
				PKP2V A 3.1s 156.0nm
				LmH B 16 1.7 μm M = 5.9
				LmV B 16 1.3 μm 5.8

Day	Phase		h m s	Remarks
18.	e	A	14 46 48.5	Turkey 37.9 N 37.9 E H = 14 41 32 h = 28 km D = 22.59 Az = 313 (ISC)
18.	e	A	17 04 06.5	Dominican Republic Region 18.82 N 69.34 W H = 16 52 43.4 h = 82 km MB = 5.1 D = 70.00 Az = 42 (ISC) traces
18.	eP	A	21 27 43.5	South Atlantic Ridge 34.18 S 20.24 W H = 21 14 50.5 h = normal MB = 5.5 D = 89.02 Az = 19.6 (NEIS)
19.	ePn	A	02 51 16.5	Switzerland 46.66 N 7.43 E H = 02 49 50.3 h = normal D = 4.86 Az = 33.3 (NEIS) PV A 1.1s 28.2nm
19.	eP	A	09 05 26	Fox Islands 52.98 N 167.96 W H = 08 53 39.2 h = 52 km MB = 5.0 D = 76.75 Az = 0 (ISC) PV A 0.8s 15.4nm M = 5.0
20.	e(pPKP)	A	02 26 24	New Britain Region 5.33 S 151.54 E H = 02 07 17.9 h = 74 km MB = 5.1 D = 123.80 Az = 330.4 (NEIS)
20.	ePKHKP	A	05 32 32.5	New Hebrides Islands 14.36 S 166.96 E
	ePKIKP	A	32 36	H = 05 13 14.8 h = normal MB=5.6 MS=4.9
	ePP	A	35 35	D = 138.66 Az = 336.3 (NEIS)
	ePKS	A	36 12	
20.	ePKIKP	A	06 13 27	Fiji Region 21.9 S 178.4 W H = 05 53 40.9 h = 0 km D = 150.26 Az = 347 (ISC)

Day	Phase		h m s	Remarks
20.	eP	A	20 16 12.5	India-East Pakistan Border Region 22.84 N 92.94 E H = 20 05 18.7 h = normal MB = 4.8 D = 67.28 Az = 317.0 (NEIS) PV A 1.2s 16.3nm M = 3.9
21.	eP	A	00 17 25	Dodecanese Islands 36.18 N 28.03 E H = 00 13 08.5 h = 77 km MB = 3.8 (ISC) D = 18.7
21.	eP	A	00 55 13	Off W. Coast of Northern Sumatra 0.55 N 96.59 E H = 00 42 32.3 h = normal MB = 5.2 D = 86.38 Az = 320.5 (NEIS)
21.	eP	A	03 56 26	Sicily 37.68 N 15.11 E H = 03 53 18.8 h = 0 km D = 13.20 Az = 350 (ISC)
21.	ePKHKP	A	07 16 28.5	Tonga Islands 18.41 S 173.87 W H = 06 56 46.5 h = 45 km MB = 4.8 D = 147.54 Az = 353.5 (NEIS)
21.	ePb	A	10 06 50	Yugoslavia 44.9 N 16.0 E H = 10 04 54 h = 0 km D = 6.49 Az = 335
21.	eP	A	14 21 54	Unimak Island Region 53.91 N 163.67 W H = 14 10 10.4 h = normal MB = 4.2 D = 75.75 Az = 3.1 (NEIS)
21.	eP	A	20 14 49	Sicily 38.94 N 15.41 E H = 20 11 57.6 h = 133 km MB = 3.8 D = 12.01 Az = 348.3 (NEIS)
21.	eP	A	20 46 50	North of Ascension Island 0.12 N 17.33 W H = 20 37 12.1 h = normal MB = 4.7 D = 56.03 Az = 21.8 (NEIS)

Day	Phase		h m s	Remarks
22.	eP	A	06 16 02	Tadzhik SSR 40.06 N 71.76 E H = 06 08 09.3 h = 50 km MB = 5.0 D = 42.47 Az = 305.1 (NEIS)
22.	e	A	08 55 31.5	
22.	+iP	ABC	13 39 40	Near East Coast of Kamchatka
	ePS	C	49 24	55.20 N 162.14 E
	eSKS	BC	49 50	H = 13 28 20.0 h = normal MB=5.7 MS=5.5
	eSS	C	54 20	D = 71.69 Az = 340.7 (NEIS)
	LmV	B	14 11.1	PV A 1.2s 148.3nm M = 5.9
	LmH	B	11.3	LmH B 22 5.8 µm 5.8
				LmV B 22 4.7 µm 5.8
22.	e(P)	A	23 18 03	Ryukyu Islands Region 24.87 N 127.05 E H = 23 05 31.7 h = 128 km MB = 4.7 D = 85.75 Az = 324.8 (NEIS)
23.	ePKIKP	A	14 10 05.5	South of Fiji Islands 22.92 S 179.10 W
	ePKHKP	AB	10 12	H = 13 51 08.8 h = 449 km MB = 5.4
	ePKP2	A	10 20.5	D = 151.09 Az = 345.8 (NEIS)
	ePKS	A	13 50	PKIKPV A 1.3s 26.2nm
				PKHKPV A 1.1 197.5nm
				PKP2V A 1.2 158.5nm
23.	ePKHKP	A	16 08 12.5	Fiji Islands Region 21.61 S 176.56 W H = 15 48 43.1 h = 191 km MB = 4.8 D = 150.32 Az = 349.5 (NEIS) PKHKPV A 1.6s 24.7nm
23.	ePKP	A	16 18 41	Fiji Islands Region 17.93 S 178.22 W H = 15 59 58.9 h = 539 km MB = 4.6 D = 146.42 Az = 348.7 (NEIS) PKPV A 1.7s 24.2nm
24.	eP	A	00 54 38	Sicily 37.9 N 14.9 E H = 00 51 31 h = 0 km D = 12.98 Az = 351 (ISC)

Day	Phase		h m s	Remarks
24.	eP	A	09 43 34.5	Greece 38.23 N 20.23 E
	LmH	B	48.5	H = 09 40 17.5 h = 56 km MB = 4.5
	LmV	B	49.5	D = 13.84 Az = 337 (ISC)
				PV A 1.0s 21.7nm M = 4.8
				LmH B 15 1.8 µm
24.	e	A	11 13 24.5	Volcano Islands Region 23.91 N 142.48 E
				H = 10 59 53.7 h = normal MB = 5.2
				D = 94.04 Az = 331.1 (NEIS)
				traces
24.	+iP	A	13 21 55.8	Tyrrhenian Sea 39.81 N 14.59 E
				H = 13 19 23.2 h = 358 km MB = 4.6
				D = 11.04 Az = 350.1 (NEIS)
				PV A 1.0s 27.6nm M = 4.6
24.	e(Pn)	A	18 23 20	Probably Czechoslovakia
	e(Sg)	A	23 40.5	
24.	eP	A	18 54 17	Southern Alaska 61.59 N 147.63 W
	epP	A	54 27	H = 18 43 26.8 h = 40 km MB = 4.8
				D = 66.89 Az = 14.2 (NEIS)
				h = 38 km
24.	+eiP	AB	19 24 50.5	Hokkaido, Japan Region
	e	A	25 03	42.13 N 143.91 E
	eS	BC	34 40	H = 19 12 52.1 h = 45 km MB=5.9 MS=6.1
	eSS	C	40 45	D = 78.66 Az = 331.3 (NEIS)
	LmH	B	20 03.1	PV A 1.6s 302.0nm M = 6.1
	LmV	B	03.1	PV B 5 2.42 µm 6.5
				LmH B 18.5 31.6 µm 6.7
				LmV B 18.5 35.5 µm 6.8
24.	eP	A	23 50 09	Hokkaido, Japan Region
	epP	A	50 23	41.93 N 143.95 E
				H = 23 38 08.5 h = 41 km MB = 5.2
				D = 78.85 Az = 331.3 (NEIS)
				h = 50 km

Day	Phase		h m s	Remarks
cont. 24.	LmH	B	24 28.5	PV A 1.5s 27.6nm M = 5.0
	LmV	B	28.5	LmH B 18 0.9 /um 5.2
				LmV B 18 1.3 /um 5.3
25.	eP	A	00 45 41	Central Italy 42.85 N 13.21 E
				H = 00 43 40.4 h = normal
				D = 7.87 Az = 352.6 (NEIS)
25.	eP	A	10 16 30	Hokkaido, Japan Region
	LmH	B	54.7	41.80 N 144.02 E
	LmV	B	55.0	H = 10 04 28.1 h = 41 km MB = 5.0
				D = 78.99 Az = 331.4 (NEIS)
				LmH B 18s 1.8 /um M = 5.4
				LmV B 18 1.7 /um 5.4
25.	ePKHKP	A	14 30 19	Fiji Islands Region 20.00 S 178.05 W
				H = 14 11 37.9 h = 606 km MB = 4.7
				D = 148.48 Az = 348.2 (NEIS)
				PKHKPV A 1.0s 23.6nm
25.	eP	A	20 41 41	Mariana Islands 18.92 N 145.49 E
	e	A	42 09	H = 20 28 13.0 h = 141 km MB = 5.9
	e	A	45 18	D = 99.74 Az = 332.2 (NEIS)
	e	A	45 32	PV A 1.5s 35.2nm M = 5.7
	ePP	AB	45 47	PPV A 2.0 308.0nm 6.4
	epPP	B	46 30	LmH B 19 3.7 /um
	eSKS	B	52 00	LmV B 16 2.5 /um
	eS	B	53 00	
	e	B	55 25	
	LmH	B	21 24.7	
	LmV	B	30.0	
25.	-iPKP	AB	22 59 57.3	Samoa Islands Region 16.43 S 172.53 W
				H = 22 40 16.1 h = 10 km MB=5.2 MS=5.0
				D = 145.71 Az = 355.3 (NEIS)
				PKPV A 1.5s 125.5nm

Day	Phase		h m s	Remarks
26.	eP	A	03 23 12	Fox Islands, Aleutian Is.
				52.30 N 171.40 W
				H = 03 11 20.9 h = 54 km MB = 5.3
				D = 77.40 Az = 358.0 (NEIS)
26.	eP	A	05 23 33.5	Turkey 37.40 N 29.76 E
	e	A	23 40.5	H = 05 19 16.6 h = 25 km MB = 4.4
				D = 18.52 Az = 321.4 (NEIS)
26.	eP	ABC	05 48 39	Near Coast of Michoacan, Mexico
	e	A	48 48	18.60 N 103.40 W
	ePP	BC	52 17	H = 05 35 33.6 h = normal MB=5.1 MS=6.1
	eSKS	BC	59 15	D = 90.63 Az = 35.2 (NEIS)
	eiSP	BC	06 00 56	PV A 1.6s 24.7nm M = 5.3
	eiSS	C	05 56	PV B 10 1.1 µm 6.2
	LmH	B	32.0	LmH B 18.5 8.3 µm 6.2
	LmV	B	34.7	LmV B 15.5 7.7 µm 6.3
27.	eP	A	07 19 00	Mariana Islands Region 21.06 N 144.53 E
				H = 07 05 41.6 h = 148 km MB = 5.3
				D = 97.44 Az = 331.9 (NEIS)
27.	eP	A	08 57 14.5	North Atlantic Ridge 33.81 N 38.56 W
	eS	BC	09 03 30	H = 08 49 41.2 h = normal MB=5.1 MS=5.3
	LmH	C	09.3	D = 39.95 Az = 49.6 (NEIS)
	LmV	B	11.2	PV A 1.7s 36.4nm M = 4.8
				LmH C 28 5.1 µm 5.2
				LmV B 18 4.1 µm 5.4
27.	eP	A	21 10 31	Crete 35.16 N 25.42 E
				H = 21 06 20.4 h = 66 km MB = 4.5
				D = 18.44 Az = 331.3 (NEIS)
28.	eP	A	03 42 42.5	Algeria 36.07 N 4.50 E
	e	A	42 46.5	H = 03 39 03.6 h = normal MB=4.8 MS=4.6
	LmH	C	48.0	D = 15.46 Az = 17.2 (NEIS)
				LmH C 18.5s 1.6 µm M = 4.2

Day	Phase		h m s	Remarks
28.	ePKP2	A	06 23 09.5	South of Fiji Islands 24.28 S 178.74 E
	e	A	23 20.5	H = 06 04 19.2 h = 585 km MB = 4.9
	epPKP	A	25 22	D = 151.85 Az = 342.5 (NEIS)
				PKP2V A 1.2s 16.3nm
				pPKPV A 1.0 11.8nm
28.	e	A	20 00 34	Central Italy 43.87 N 10.74 E
				H = 19 57 24.3 h = 6 km (ISC)
				D = 6.8
29.	e(P)	A	04 25 16	Near East Coast of Kamchatka
				55.36 N 161.94 E
				H = 04 13 51.2 h = 0 km MB = 4.5
				D = 71.51 Az = 341 (ISC)
29.	eP	A	13 37 08.5	Sicily 37.7 N 15.05 E
	LmH	C	41.0	H = 13 34 02 h = 21 km
				D = 13.19 Az = 350 (ISC)
				LmH C 18s 1.6/um M = 4.1
29.	e(PKP2)	A	14 59 14.5	Tonga Islands 20.35 S 173.77 W
				H = 14 39 15.0 h = normal MB=4.8 MS=5.0
				D = 149.47 Az = 353.2 (NEIS)
				(PKP2)V A 1.8s 37.2nm
29.	eP	A	15 16 09	Greece 38.39 N 21.80 E
	eX	A	16 11	H = 15 12 44.9 h = 31 km MB = 4.4
				D = 14.22 Az = 332.7 (NEIS)
				XV A 1.2s 12.2nm
29.	ePKP	A	19 15 32.5	Banda Sea 7.42 S 128.58 E
	e	A	16 18	H = 18 57 13.1 h = 154 km MB = 5.7
				D = 112.66 Az = 322.1 (NEIS)
				PKPV A 1.0s 21.7nm
29.	ePKP	A	22 55 45	Banda Sea 7.26 S 128.54 E
	e	A	56 08	H = 22 37 24.6 h = 154 km MB = 5.4
	e	A	56 29.5	D = 112.51 Az = 322.1 (NEIS)

Day	Phase		h m s	Remarks
30.	eP1	A	05 04 46.5	Eastern Kazakh SSR 49.84 N 78.08 E
	+iP2	A	04 50.8	H = 04 57 02.1 h = 0 km MB = 5.4
	eP2n	A	06 23	D = 41.21 Az = 297.6 (NEIS)
				Underground explosion MB = 6.0 (UPP)
				P2V A 0.8s 127.0nm M = 5.7
				P2nV A 0.9 21.4nm
30.	ePKIKP	A	10 11 50	Aroe Islands Region 5.16 S 134.07 E
	e	A	12 40	H = 09 53 12.0 h = normal MB=5.9 MS=6.3
	ePP	AC	12 47	D = 114.15 Az = 323.9 (NEIS)
	e	A	12 52	PKIKPV A 1.5s 25.2nm
	ePS	C	22 16	LmH B 18 8.3/um M = 6.4
	ePPS	C	23 28	LmV B 18 8.0/um 6.4
	eSS	C	28 40	
	LmV	B	11 06.5	
	LmH	B	06.7	
30.	e	A	22 59 32	Hungary 47.2 N 18.7 E
				H = 22 56 46 h = 0 km (ISC)
				D = 5.8
31.	-eP	AB	07 16 18	Kyushu, Japan 31.84 N 131.64 E
	e	A	16 30	H = 07 03 58.1 h = 37 km MB=5.6 MS=5.4
	ePS	C	26 44	D = 82.32 Az = 326.2 (NEIS)
	eSS	C	32 30	PV A 2.0s 172.0nm M = 5.8
	LmV	B	57.0	LmH B 16.5 17.1/um 6.5
	LmH	B	57.2	LmV B 16 21.6/um 6.7
31.	ePKP	A	15 29 04	Fiji Islands Region 17.76 S 178.67 W
	e	A	29 07	H = 15 10 29.1 h = 584 km MB = 5.3
				D = 146.18 Az = 348.2 (NEIS)
31.	+eiP	AB	20 07 19	Fox Islands, Aleutian Is.
				52.36 N 168.74 W
				H = 19 55 26.2 h = 36 km MB=5.6 MS=5.0
				D = 77.37 Az = 359.8 (NEIS)
				PV A 1.3s 178.0nm M = 5.9

Day	Phase		h m s	Remarks
31.	-eP	A	20 27 47.5	Fox Islands, Aleutian Is.
				52.25 N 168.78 W
				H = 20 15 54.6 h = 44 km MB = 4.8
				D = 77.49 Az = 359.7 (NEIS)
				PV A 1.0s 25.6nm M = 5.2
31.	ePKIKP	A	20 35 24	Solomon Islands 7.50 S 155.98 E
	LmH	B	21 23.0	H = 20 16 22.5 h = 62 km MB = 5.3
	LmV	B	36.0	D = 127.81 Az = 332.0 (NEIS)
				LmH B 25s 3.3/um
31.	ePKIKP	AB	23 49 09	Solomon Islands 7.46 S 155.89 E
	eX	A	49 25	H = 23 30 05.3 h = 34 km MB=6.0 MS=7.0
	ePP	B	51 15	D = 127.74 Az = 332.0 (NEIS)
	eSS	B	24 08 30	PKIKPV A 1.6s 88.0nm
	e	B	11 48	XV A 2.0 274.0nm
	LmH	B	50.7	LmH B 18 41.3/um M = 7.2
	LmV	B	50.9	LmV B 18 49.5/um 7.2
31.	ePKPKP	A	23 57 08	Solomon Islands 7.00 S 155.78 E
				H = 23 38 03.9 h = 33 km MB = 5.8
				D = 127.28 Az = 332 (ISC)

Day	Phase		h m s	Remarks
1.	e	A	00 02 26.5	
1.	eP	A	00 04 50	Turkey 38.55 N 27.02 E
	i	A	04 56	H = 00 01 02.4 h = 29 km MB = 5.2
	i	A	05 00	D = 16.28 Az = 322.9 (NEIS)
	ei	AB	05 07	
1.	ePKIKP	A	01 23 16	Solomon Islands 7.31 S 155.90 E
				H = 01 04 14.5 h = 49 km MB = 5.2
				D = 127.61 Az = 332.0 (NEIS)
				PKIKPV A 1.3s 26.2nm
1.	ePKIKP	A	03 31 36	Solomon Islands 7.38 S 155.58 E
	eiX	AB	31 46	H = 03 12 33.1 h = 40 km MB=6.2 MS=7.1
	+eiPP	B	33 44	D = 127.52 Az = 331.8 (NEIS)
	ePPS	B	45 55	PKIKPV A 1.5s 37.7nm
	eSS	B	51 30	XV A 2.0 47.9nm
	LmH	B	04 16.0	PPV B 11 5.6/um M = 6.7
	LmV	B	31.5	LmH B 30 133.3/um 7.4
				LmV B 20 55.8/um 7.2
1.	ePKIKP	A	04 38 24	Solomon Islands 7.22 S 155.97 E
				H = 04 19 20.6 h = normal MB = 5.0
				D = 127.56 Az = 332.1 (NEIS)
1.	ePKIKP	A	05 31 16	Solomon Islands 7.17 S 155.62 E
				H = 05 12 13.9 h = normal MB = 5.0
				D = 127.35 Az = 331.9 (NEIS)
1.	ePKP	A	05 33 33.5	Tonga 17.12 S 175.08 W
				H = 05 14 23.0 h = 266 km MB = 4.7
				D = 146.13 Az = 352 (ISC)
1.	e(P)	A	05 54 43	
1.	ePKIKP	A	07 21 22	Solomon Islands 7.59 S 156.03 E
				H = 07 02 14.3 h = normal MB=5.4 MS=5.6
				D = 127.91 Az = 332.0 (NEIS)

Day	Phase		h m s	Remarks
1.	e	A	08 31 23	Solomon Islands 7.23 S 155.23 E H = 08 12 18.9 h = 83 km MB = 5.5 D = 127.23 Az = 322 (ISC)
1.	ePKIKP e	A A	08 35 11.5 35 20	Solomon Islands 7.77 S 155.60 E H = 08 16 09.9 h = normal MB = 5.3 D = 127.87 Az = 331.7 (NEIS)
1.	-ePKIKP	A	09 27 26.5	Solomon Islands 7.21 S 155.79 E H = 09 08 24.5 h = 48 km MB = 5.3 D = 127.46 Az = 332.0 (NEIS) PKIKPV A 1.5s 47.7nm
1.	ePKIKP	A	11 07 51	Solomon Islands 7.28 S 155.92 E H = 10 48 47.6 h = normal MB = 5.3 D = 127.59 Az = 332.0 (NEIS)
1.	eP	A	12 14 04.5	Southern Sumatra 4.63 S 103.42 E H = 12 00 57.1 h = 136 km MB = 5.2 D = 94.69 Az = 320.3 (NEIS) PV A 1.2s 16.3nm M = 5.2
1.	+eP e e LmH	A A A C	15 16 13.5 16 22 16 28 16 26.0	Near East Coast of Kamchatka 54.42N 162.19E H = 15 04 48.9 h = 30 km MB=5.1 MS=5.4 D = 72.44 Az = 340.8 (NEIS) PV A 1.8s 74.4nm M = 5.4 LmH C 30 1.4/um 5.1
1.	eSg	A	19 56 26	Northern Italy 47.0 N 10.88 E H = 19 54 31.2 h = 0 km D = 3.69 Az = 7 (ISC)
1.	ePKIKP	A	23 35 59	Solomon Islands 7.24 S 155.43 E H = 23 16 55.2 h = 47 km MB = 5.4 D = 127.33 Az = 331.8 (NEIS) traces

Day	Phase		h m s	Remarks
2.	eP	A	03 44 27	Azores Islands Region 35.63 N 34.54 W
	LmH	C	54.0	H = 03 37 25.0 h = normal MB = 4.9
	LmV	C	57.0	D = 36.25 Az = 51.0 (NEIS)
				PV A 1.6s 32.9nm M = 5.0
				LmH C 24 0.7 µm 4.3
				LmV C 17 1.1 µm 4.8
2.	+iPKP	AB	08 46 43	New Hebrides Islands 19.10 S 169.49 E
	e	A	47 37	H = 08 27 40.2 h = 269 km MB = 5.6
				D = 143.96 Az = 336.0 (NEIS)
				PKPV A 1.1s 117.0nm
2.	ePKP	A	12 03 30	Aroe Islands Region 5.02 S 134.01 E
	LmH	B	48.7	H = 11 44 52.9 h = normal MB=5.6 MS=5.8
	LmV	B	56.4	D = 114.01 Az = 323.9 (NEIS)
				LmH B 18s 4.6 µm M = 6.1
				LmV B 18 2.2 µm 5.8
2.	ePKP	A	12 38 05	Tonga Islands 17.42 S 174.06 W
				H = 12 18 30.0 h = 73 km MB = 5.1
				D = 146.54 Az = 353.4 (NEIS)
2.	eiP	AB	16 06 17	Southern Alaska 61.60 N 147.60 W
	eipP	A	16 28	H = 15 55 28.3 h = 48 km MB=5.1 MS=4.7
	LmH	C	37.0	D = 66.88 Az = 14.2 (NEIS)
	LmV	C	51.3	h = 44 km
				PV A 1.5s 50.2nm
				LmH C 37.5 4.6 µm M = 5.3
				LmV C 21 1.9 µm 5.4
2.	e	A	16 34 39	
2.	eP	A	20 09 42	Sunda Strait 6.22 S 104.26 E
	eP	A	09 53	H = 19 56 11.4 h = normal MB=5.2 MS=5.6
	ePP	A	13 45.5	D = 96.44 Az = 320.2 (NEIS)
	epPP	A	13 55	h = 43 km
	LmV	B	21 02.0	PV A 1.5s 45.2nm M = 5.7
	LmH	B	05.0	PPV A 1.5 32.6nm 5.6
				LmH B 20 2.2 µm 5.7
				LmV B 16 1.3 µm 5.5

Day	Phase		h m s	Remarks
3.	eP	A	06 36 22.5	Arabian Sea 14.50 N 56.09 E H = 06 27 23.2 h = normal MB = 4.8 D = 50.82 Az = 324.9 (NEIS)
3.	eP	A	09 03 12.5	Arabian Sea 14.52 N 55.74 E H = 08 54 09.1 h = normal MB = 5.0 D = 50.60 Az = 325.0 (NEIS) (P)V A 1.0s 15.7nm M = 5.0
3.	-iP	A	10 21 31	Luzon, Philippine Islands
	epP	AB	21 41	18.91 N 120.07 E
	esP	A	21 46	H = 10 08 48.4 h = 30 km MB=5.9 MS=5.2
	e	A	22 03.5	D = 86.70 Az = 322.8 (NEIS)
	ePP	A	25 00	h = 39 km
	eSKS	B	31 50	PV A 1.6s 82.6nm M = 5.7
	ePS	B	33 00	LmH B 15 4.5 μm 6.0
	eSS	C	37 50	LmV B 13.5 4.5 μm 6.1
	LmH	B	11 04.6	
	LmV	B	06.3	
3.	ePKIKP	A	16 32 02.5	Solomon Islands 7.33 S 155.50 E
	LmH	B	17 29.6	H = 16 12 56.7 h = 43 km MB=5.4 MS=5.6
	LmV	B	33.4	D = 127.44 Az = 331.8 (NEIS)
				PKIKPV A 2.0s 51.2nm
				LmH B 18 1.4 μm M = 5.7
				LmV B 18 1.3 μm 5.6
3.	ePKP	A	19 04 24	Aroe Islands Region 5.13 S 133.85 E
	LmH	B	49.6	H = 18 45 45.7 h = normal MB=5.7 MS=5.2
	LmV	C	57.5	D = 114.00 Az = 323.9 (NEIS)
				LmH B 18s 1.4 μm M = 5.6
				LmV B 20 1.1 μm 5.4
4.	eP	A	03 35 23	Hokkaido, Japan Region 43.67 N 140.82 E H = 03 23 54.9 h = 218 km MB = 4.7 D = 76.20 Az = 329.5 (NEIS)

Day	Phase		h m s	Remarks
4.	ePn	A	14 03 03.5	Austria 47.63 N 14.05 E
				H = 14 02 08.4 h = 5 km
				D = 3.42 Az = 333 (ISC)
4.	ePKIKP	AB	20 29 45	Solomon Islands 7.30 S 155.83 E
	e	A	29 49	H = 20 10 42.0 h = 55 km MB = 5.4
	e	A	29 53	D = 127.56 Az = 332.0 (NEIS)
	ePP	BC	31 40	PKIKPV A 1.5s 35.2nm
	ePKS	BC	33 00	LmH B 22 5.4/um M = 6.2
	e(PPP)	C	34 40	LmV B 19 5.3/um 6.2
	ePS	C	41 44	
	ePPS	C	43 25	
	eSS	C	48 32	
	LmH	B	21 17.3	
	LmV	B	30.0	
4.	e	A	21 44 07	Solomon Islands 7.26 S 155.94 E
				H = 21 24 56.4 h = 53 km MB = 5.0
				D = 127.58 Az = 332.1 (NEIS)
				traces
4.	eP	A	23 18 06.5	Tyrrhenian Sea 39.3 N 12.7 E
				H = 23 15 20 h = 0 km
				D = 11.37 Az = 356 (ISC)
4.	e	A	24 03 56	Solomon Islands 7.33 S 155.74 E
				H = 23 44 44.6 h = 52 km MB = 5.2
				D = 127.56 Az = 331.9 (NEIS)
5.	e(P)	A	02 31 00.5	Turkey 39.79 N 26.79 E
				H = 02 27 16.7 h = 0 km
				D = 15.21 Az = 321 (ISC)
5.	epP	A	02 36 19.5	Central Alaska 62.70 N 148.85 W
				H = 02 25 22.0 h = 75 km MB = 5.0 (NEIS)
				D = 66.0
				PV A 0.7s 19.2nm

Day	Phase		h m s	Remarks
5.	eP	A	06 45 39	Algeria 35.98 N 4.39 E H = 06 41 59.5 h = normal MB = 4.4 D = 15.56 Az = 17.4 (NEIS)
5.	eP	AB	15 09 22	Dodecanese Islands 36.82 N 27.02 E H = 15 05 24.2 h = 153 km MB = 4.8 D = 17.69 Az = 326.2 (NEIS) PV A 1.9s 121.1nm M = 4.9
5.	e	A	15 27 23	Poland 50.6 N 19.5 E H = 15 25 52 h = 0 km D = 5.02 Az = 274 (ISC)
5.	eP	A	18 27 43	Turkey 37.31 N 29.58 E H = 18 23 26.2 h = 39 km MB = 4.4 D = 18.51 Az = 321.8 (NEIS) PV A 1.5s 38.7nm M = 4.4
5.	ePKP	A	22 47 01.5	Fiji Islands Region 16.18 S 177.75 W
	LmH	C	23 40.0	H = 22 27 26.3 h = 62 km MB = 5.0 D = 144.80 Az = 349.6 (NEIS) PKPV A 1.4s 16.3nm
6.	eP	A	03 02 56	Off East Coast of Kamchatka 52.67 N 160.35 E H = 02 51 22.1 h = normal MB = 4.7 D = 73.73 Az = 339.9 (NEIS)
6.	-iP1	AB	04 15 57.3	Unimak Island Region 53.80 N 164.67 W
	-iP2	A	16 00	H = 04 04 07.2 h = 2 km MB=5.9 MS=6.5
	iP3	A	16 06	D = 75.88 Az = 2.4 (NEIS)
	ePP	C	18 58	P1V A 1.3s 183.5nm M = 5.9
	ePPP	C	20 30	P1V B 8 5.2/um 6.6
	eS	BC	25 40	P2V A 1.8 337.8nm 6.1
	ePS	C	26 25	P3V A 1.9 447.0nm 6.2
	ePPS	C	26 50	LmH B 18 16.7/um 6.4
	eSS	C	30 48	LmV B 17 13.4/um 6.4
	eSSS	C	33 50	

Day	Phase		h m s	Remarks
6.	iP	A	17 34 09.2	Afghanistan-USSR Border Region
	LmH	C	18 07.0	36.52 N 71.42 E
				H = 17 26 16.5 h = 193 km MB = 4.9
				D = 44.37 Az = 308.1 (NEIS)
				LmH C 20s 1.0/um
7.	eP	A	01 30 18	South Atlantic Ridge 12.35 S 14.78 W
				H = 01 19 26.9 h = normal MB = 4.7
				D = 66.83 Az = 17.9 (NEIS)
				PV A 1.9s 45.5nm M = 5.2
7.	ePKP	A	08 03 49	Tonga Islands 19.05 S 174.14 W
				H = 07 44 05.1 h = normal MB = 4.8
				D = 148.15 Az = 353.0 (NEIS)
7.	e(P)	A	08 50 33	Turkey 39.70 W 26.88 E
				H = 08 46 51.9 h = 37 km MB = 4.2
				D = 15.31 Az = 321 (ISC)
7.	LmH	B	10 44.6	West Irian Region 3.79 S 134.01 E
	LmV	B	54.5	H = 09 41 48 h = 10 km MB = 5.8 (ISC)
				D = 113.0
				LmH B 17s 1.9/um M = 5.7
7.	i(P)	A	19 15 51.3	Kurile Islands 49.86 N 156.56 E
	LmH	B	53.4	H = 19 04 07.0 h = 49 km M = 4.9
	LmV	B	54.8	D = 75.49 Az = 338 (ISC)
				LmH B 18s 0.7/um M = 5.0
7.	e	A	20 25 52	
	e	A	26 00	
8.	ePKHKP	A	00 53 24	Tonga Islands 20.34 S 173.81 W
	ePKP2	A	53 39	H = 00 33 35.9 h = 37 km MB = 4.9
				D = 149.46 Az = 353.2 (NEIS)
8.	e	A	03 18 28	Mid-Indian Rise 9.22 S 66.03 E
				H = 03 06 29.5 h = 33 km MB = 4.5
				D = 76.41 Az = 327 (ISC)

Day	Phase		h m s	Remarks
8.	+iP	A	14 33 08.0	Komandorsky Islands Region
	e	A	33 16.5	54.40 N 167.60 E
	ePcP	A	33 26	H = 14 21 37.4 h = normal MB=5.4 MS=5.0
	LmH	B	15 10.7	D = 73.41 Az = 344.3 (NEIS)
	LmV	B	11.8	PV A 0.8s 53.8nm M = 5.6
				LmH B 17 0.9 µm 5.1
				LmV B 16 0.8 µm 5.1
8.	ePKP	A	18 44 10	Loyalty Islands Region
	LmH	C	19 32.8	21.38 S 170.10 E
				H = 18 24 32.2 h = normal MB=5.4 MS=5.8
				D = 146.25 Az = 335.1 (NEIS)
				PKPV A 1.3s 15.3nm
				LmH C 38 2.1 µm M = 5.6
8.	e	A	20 14(15)	France 44.22 N 6.54 E
	e	A	14 35	H = 20 12 17.7 h = 33 km
				D = 7.29 Az = 26 (ISC)
9.	e	A	03 27 43	
9.	ePKP	A	03 56 25	Tonga Islands Region 18.07 S 172.83 W
				H = 03 36 42.1 h = normal MB = 4.6
				D = 147.31 Az = 354.8 (NEIS)
				PKPV A 1.2s 14.3nm
9.	eP	A	08 41 04.5	Luzon, Philippine Islands
				15.93 N 119.77 E
				H = 08 28 15.3 h = 65 km MB = 5.2
				D = 88.88 Az = 322.8 (NEIS)
				PV A 1.2s 18.3nm M = 5.2
9.	eP	A	18 33 51	Kurile Islands 50.61 N 157.24 E
				H = 18 22 14.8 h = 57 km MB = 4.9
				D = 74.95 Az = 338.1 (NEIS)

Day	Phase		h m s	Remarks
10.	eP	A	09 03 56	Southern Honshu, Japan 35.11 N 136.92 E H = 08 51 41.2 h = 55 km MB = 5.0 D = 81.96 Az = 328.3 (NEIS)
10.	e(Pn)	A	22 35 33	Switzerland 46.6 N 6.9 E
	e(Sn)	A	36 37.5	H = 22 34 02 h = 0 km D = 5.13 Az = 36 (ISC)
11.	eP	A	01 54 13	Sunda Strait 6.10 S 104.11 E H = 01 40 46.2 h = normal MB=5.5 MS=5.2 D = 96.25 Az = 320.2 (NEIS) PV A 1.3s 19.7nm M = 5.4
11.	eP	A	14 22 14.5	Nicobar Islands Region 6.91 N 91.34 E H = 14 10 12.6 h = 7 km MB = 5.2 D = 78.17 Az = 320.2 (NEIS)
12.	+eP	A	10 00 44	Mindoro, Philippine Islands 13.52 N 120.46 E H = 09 47 44.8 h = 74 km MB = 5.5 D = 91.19 Az = 322.9 (NEIS) PV A 1.5s 50.3nm M = 5.6
12.	ePKP	A	21 49 39	Loyalty Islands 20.92 S 168.82 E H = 21 29 59.1 h = normal MB = 4.6 D = 145.33 Az = 334.3 (NEIS)
13.	ePKHKP	A	00 33 27	Tonga Islands Region 22.38 S 174.62 W
	e(PKP2)	A	33 39	H = 00 13 34.3 h = normal MB = 4.6 D = 151.36 Az = 351.7 (NEIS) PKHKPV A 1.2s 26.4nm
13.	eP	A	23 51 54	Northern Celebes 0.01 N 122.74 E H = 23 37 52.9 h = 11 km MB=5.8 MS=4.8 D = 103.26 Az = 322.4 (NEIS)

Day	Phase		h m s	Remarks
14.	ePKIKP	A	02 27 02.5	South of Fiji Islands 22.3 S 179.4 W H = 02 07 21.8 h = 33 km MB = 4.9 D = 150.38 Az = 346 (ISC)
14.	eP	A	06 50 46.5	Northern Sumatra 2.55 N 98.97 E H = 06 38 06.5 h = 34 km MB=5.6 MS=5.0 D = 86.37 Az = 320.4 (NEIS) PV A 1.9s 75.7nm M = 5.6
14.	eP	A	12 10 13	North Atlantic Ridge 21.98 N 44.25 W H = 12 01 07.0 h = normal MB = 5.4 D = 51.79 Az = 42.1 (NEIS) PV A 1.2s 20.3nm M = 5.0
14.	eP	A	14 58 06.5	Norwegian Sea 72.32 N 1.17 E H = 14 53 09.7 h = normal MB = 4.6 D = 22.25 Az = 162.2 (NEIS) PV A 1.3s 17.5nm M = 4.3
15.	e(P)	A	04 04 35.5	Philippine Islands Region 20.38 N 121.55 E H = 03 51 46.6 h = 36 km MB = 5.1 D = 86.37 Az = 323.1 (NEIS) (P)V A 1.6s 27.5nm (M = 5.2)
15.	LmH	B	09 11.3	Ryukyu Islands 28.75 N 128.53 E H = 08 24 01.7 h = 48 km MB = 4.7 (ISC) D = 83.2 LmH B 18s 2.7/um M = 5.7
16.	eP e LmH LmV	A A B B	02 02 51.5 05 27.5 42.5 44.9	Andaman Islands Region 11.43 N 92.35 E H = 01 51 10.8 h = 25 km MB=5.5 MS=6.0 D = 75.38 Az = 319.5 (NEIS) PV A 1.8s 33.8nm M = 5.1 LmH B 18.5 5.5/um 5.9 LmV B 18 4.0/um 5.8

Day	Phase		h m s	Remarks
16.	ePKP2 epPKP	A A	05 59 35 06 01 39.5	<u>Kermadec Islands Region</u> 31.54 S 179.16 E H = 05 39 56.2 h = 499 km MB = 5.3 D = 158.78 Az = 338 (ISC) PKP2V A 1.5s 95.5nm
16.	ePKP	A	07 16 55.5	<u>Loyalty Islands Region</u> 22.41 S 171.31 E H = 06 57 28.5 h = 119 km MB = 5.1 D = 147.65 Az = 335.6 (NEIS) PKPV A 2.5s 76.8nm
16.	eP	A	16 22 05.5	<u>Taiwan Region</u> 21.54 N 121.54 E H = 16 09 38.1 h = 115 km MB = 4.7 D = 85.44 Az = 323.1 (NEIS)
17.	eP epP esP	A A A	04 46 55 47 11 47 15.5	<u>Mariana Islands</u> 18.44 N 146.69 E H = 04 33 11.9 h = 58 km MB = 5.6 D = 100.69 Az = 332.8 (NEIS) h = 57 km
17.	ePP	A	05 06 48	<u>Greece</u> 38.01 N 21.71 E H = 05 03 07.4 h = normal MB = 3.9 D = 14.53 Az = 333.6 (NEIS)
17.	e(P)	A	11 41 55	<u>Ionian Sea</u> 37.3 N 20.9 E H = 11 38 19 h = 33 km MB = 3.9 D = 14.87 Az = 336 (ISC)
17.	eP	A	12 18 43	<u>Southern Greece</u> 37.11 N 21.14 E H = 12 15 07.1 h = 0 km MB = 4.1 D = 15.15 Az = 340 (ISC)
17.	eP	A	21 31 38.5	<u>Unimak Island Region</u> 53.65 N 163.54 W H = 21 19 53.2 h = normal MB = 4.7 D = 76.00 Az = 3.2 (NEIS) PV A 1.1s 16.1nm M = 4.9
18.	ePKHKP	A	09 11 24	<u>South of Fiji Islands</u> 23.51 S 179.83 E H = 08 52 30.5 h = 256 km MB = 4.5 D = 151.40 Az = 344 (ISC)

Day	Phase		h m s	Remarks
18.	eP	A	21 13 33.5	Off East Coast of Honshu, Japan
	ePcP	A	13 42	39.71 N 143.53 E
	LmH	C	48.2	H = 21 01 22.3 h = normal MB=4.7 MS=5.3
	LmV	B	55.6	D = 80.64 Az = 331.3 (NEIS)
				PV A 1.9s 45.5nm M = 5.1
				LmH C 19 1.5/um 5.4
				LmV B 16 1.1/um 5.3
19.	e(PKP)	A	01 42 29	Timor 8.98 S 123.95 E
	e	A	43 04	H = 01 23 49.2 h = normal MB=5.3 MS=5.3
				D = 111.01 Az = 320.9 (NEIS)
				(PKPV) A 3.0s 79.0nm
19.	eP	A	02 15 19.5	East of Lake Baikal 55.0 N 113.4 E
				H = 02 05 40 h = 0 km MB = 4.2
				D = 55.86 Az = 311 (ISC)
19.	eP	AB	03 43 30	Luzon, Philippine Islands
	eSKS	C	54 28	13.91 N 122.12 E
	ePPS	B	56 32	H = 03 30 21.8 h = 17 km MB=5.7 MS=6.1
	LmH	C	04 19.9	D = 91.85 Az = 323.4 (NEIS)
	LmV	B	28.8	PV A 1.8s 128.0nm M = 6.0
				LmH C 25.3 48.2/um 6.8
				LmV B 17.5 15.7/um 6.5
19.	eP	A	04 17 06	Hindu Kush Region 35.96 N 70.87 E
				H = 04 09 02.9 h = 85 km MB = 5.0
				D = 44.36 Az = 308.5 (NEIS)
20.	ePKHKP	A	00 56 22.5	Tonga Islands 20.78 S 174.89 W
				H = 00 36 37.7 h = 60 km MB = 5.0
				D = 149.76 Az = 351.8 (NEIS)
				PKHKPV A 1.6s 27.5nm
20.	eP	A	11 51 03	Kirgiz SSR 40.66 N 73.22 E
	LmH	C	12 10.5	H = 11 43 03.9 h = 25 km MB = 5.1
	LmV	B	10.6	D = 43.05 Az = 304.9 (NEIS)
				LmV B 12s 1.3/um M = 5.1

Day	Phase		h m s	Remarks
20.	eP	A	16 22 41	Dominican Republic Region
	LmH	C	49.0	19.58 N 70.02 W
	LmV	B	51.8	H = 16 11 26.8 h = 18 km MB = 4.9
				D = 69.87 Az = 42.1 (NEIS)
				LmH C 19s 1.3 μm M = 5.2
				LmV B 16 0.9 μm 5.2
21.	e(Pg)	A	04 25 06.5	Upper Silesia, Poland
	e	A	25 30	Rockburst (WAR)
	e(Sg)	A	26 16	
21.	ePKP	A	17 39 35.5	Banda Sea 7.58 S 127.41 E
				H = 17 21 17.8 h = 165 km MB = 5.4
				D = 112.07 Az = 321.8 (NEIS)
22.	+eP1	A	00 48 39.5	Near S. Coast of Southern Honshu
	-iP2	AB	48 42.2	33.15 N 136.91 E
	Pm	A	48 47	H = 00 36 53.8 h = 385 km MB = 6.0
	epP	AB	50 10	D = 83.62 Az = 328.5 (NEIS)
	esP	B	50 50	h = 395 km
	ePP	A	51 59	P1V A 1.4s 46.6nm
	iSKS	BC	58 24.5	P2V A 1.2 187.0nm
	eiS	AB	58 26	PmV A 1.3 340.0nm M = 5.9
	eSP	B	59 10	SH B 8.5 31.2 μm 6.9
	ePS	C	01 00 04	P'P'V A 1.8 60.8nm
	eSS	C	03 50	LmH C 16 7.1 μm
	esSS	C	06 25	LmV C 16 6.0 μm
	ePKKP	A	06 57.5	
	e	C	07 14	
	e	C	10 07	
	eP'P'	A	14 57.5	
	eSKPP'	A	17 45	
	LmH	C	33.2	
	LmV	C	33.2	

Day	Phase		h m s	Remarks
22.	+iP	AB	03 41 27.7	Afghanistan-USSR Border Region
	e	A	42 00	36.50 N 71.50 E
	ePPP	A	43 47	H = 03 33 26.5 h = 116 km MB = 5.4
	eSSS	AC	52 10.5	D = 44.43 Az = 308.1 (NEIS)
				PV A 2.0s 222.0nm M = 5.6
22.	+eP	AB	07 14 57.5	West Pakistan 29.69 N 67.60 E
	LmH	B	41.7	H = 07 06 32.6 h = normal MB = 5.0
	LmV	B	41.7	D = 46.38 Az = 313.2 (NEIS)
				PV A 1.2s 56.9nm M = 5.4
				LmV B 13 0.9/um 5.0
22.	eP	A	13 43(25)	Rumania 45.65 N 26.31 E
	e	A	43 33	H = 13 40 49.0 h = 149 km MB = 4.2
				D = 11.01 Az = 302.3 (NEIS)
23.	e(P)	A	01 32 19.5	Greece 38.06 N 21.77 E
				H = 01 28 45.5 h = 44 km MB = 4.4
				D = 14.51 Az = 333.4 (NEIS)
23.	eP	A	04 26 51	Hokkaido, Japan Region 42.22 N 143.05 E
	esP	A	27 13.5	H = 04 14 56.7 h = 64 km MB = 5.3
				D = 78.27 Az = 330.8 (NEIS)
				h = 61 km
				PV A 0.8s 25.5nm M = 5.2
23.	ePKP	A	07 29 55	Fiji Region 17.7 S 177.91 W
				H = 07 10 18.4 h = 0 km MB = 4.6
				D = 146.26 Az = 349 (ISC)
				traces
23.	eP	A	20 58 33	Greece 38.35 N 20.22 E
				H = 20 55 19.1 h = normal MB = 4.4
				D = 13.73 Az = 336.4 (NEIS)
25.	eP	A	01 43 01	Ascension Islands Region
	LmH	B	02 06.0	11.52 S 13.30 W

Day	Phase		h m s	Remarks
cont. 25.	LmV	B	02 10.4	H = 01 32 17.3 h = normal MB=5.0 MS=4.9
				D = 65.61 Az = 17.1 (NEIS)
				PV A 1.5s 20.1nm M = 5.0
				LmH B 21 0.9 /um 5.0
				LmV B 16 0.7 /um 5.0
25.	+iP	AB	05 58 26.2	Kurile Islands 44.04 N 147.80 E
	LmH	B	06 36.4	H = 05 46 25.1 h = 12 km MB=5.9 MS=5.4
	LmV	B	37.8	D = 78.29 Az = 333.2 (NEIS)
				PV A 1.1s 214.0nm M = 6.1
				LmH B 21 5.2 /um 5.8
				LmV B 15.5 2.8 /um 5.7
25.	e	A	16 18 23	
25.	+iPn	A	20 05 58.2	United Kingdom 51.58 N 3.04 W
				H = 20 03 44.1 h = normal MB = 4.3
				D = 9.27 Az = 90.0 (NEIS)
26.	e	A	00 19 02	
26.	ePKP	A	00 21 32	Fiji Islands Region 16.8 S 178.3 W
				H = 00 01 57.6 h = 0 km MB = 4.9
				D = 145.36 Az = 349 (ISC)
26.	+eP	AB	06 35 11.5	Near East Coast of Kamchatka
	esP	A	35 30	53.32 N 159.70 E
				H = 06 23 45.3 h = 49 km MB=5.6 MS=4.7
				D = 72.98 Az = 339.4 (NEIS)
				h = 50 km
26.	LmV	B	11 04.5	South of the Marianas 12.04 N 143.76 E
	LmH	B	04.6	H = 09 55 19.6 h = 33 km MB = 5.2
				or 12.11 N 143.69 E
				H = 09 58 39.9 h = 42 km MB = 5.5 (ISC)
				D = 105.0
				LmH B 17s 0.9 /um M = 5.4
				LmV B 18 1.2 /um 5.7

Day	Phase		h m s	Remarks
26.	eP	A	11 16 50	Ionian Sea 37.24 N 20.85 E
	LmH	B	22.4	H = 11 13 20.7 h = normal MB = 4.4
	LmV	B	23.8	D = 14.94 Az = 336.7 (NEIS)
				LmH B 12s 0.8 µm M = 4.1
				LmV B 12 0.9 µm 4.2
26.	e	A	14 23 32.5	Mediterranean Sea 35.18 N 22.90 E
				H = 14 19 19.8 h = 47 km MB = 4.2
				D = 17.50 Az = 336 (ISC)
				traces
26.	e(P)	A	23 28 34	Off Coast of Oregon 44.01 N 128.3 W
				H = 23 16 24.3 h = 33 km MB = 4.6
				D = 79.48 Az = 25 (ISC)
27.	eP	A	03 54 09	Off Coast of Oregon 43.88 N 128.44 W
				H = 03 42 01.4 h = 15 km
				MB = 5.0 MS = 4.9 (NEIS)
				D = 80.0
				PV A 1.5s 15.1nm M = 4.8
27.	eP	A	03 55 24	Off Coast of Oregon 43.85 N 128.55 W
				H = 03 43 19.2 h = normal MB = 5.0
				D = 79.71 Az = 24.5 (NEIS)
				PV A 1.5s 35.2nm M = 5.1
27.	eP	A	04 39 13	Ascension Island Region
	e	A	39 19	11.60 S 13.41 W
	eX	A	39 25	H = 04 28 30.3 h = normal MB=5.1 MS=4.7
	LmH	B	05 02.2	D = 65.72 Az = 17.2 (NEIS)
	LmV	B	08.8	PV A 1.9s 30.5nm M = 5.1
				XV A 1.9 53.0nm
				LmH B 20.8 1.5 µm 5.2
				LmV B 16 1.4 µm 5.3

Day	Phase		h m s	Remarks
27.	eP	A	17 12 18	Southern Nevada 37.10 N 116.05 W H = 17 00 00.1 h = 0 km MB = 5.8 D = 81.25 Az = 30.7 (NEIS) Nuclear explosion LATIR (AEC) PV A 1.2s 40.7nm M = 5.4
27.	-iP	A	18 14 30	Northern Sumatra 1.27 N 97.67 E
	e	A	14 39	H = 18 01 48.7 h = normal MB=5.9 MS=5.4
	LmH	B	56.2	D = 86.51 Az = 320.5 (NEIS)
	LmV	B	19 03.3	PV A 1.3s 155.0nm M = 6.1 LmH B 20 0.6 µm 5.0 LmV B 16.5 0.6 µm 5.1
27.	ePKP	A	19 12 42.5	Samoa Islands Region 16.59 S 172.30 W H = 18 53 04.4 h = normal MB = 5.1 D = 145.88 Az = 355.6 (NEIS) PKPV A 1.4s 51.1nm
27.	ePKHKP	ABC	20 57 36	Tonga Islands 17.85 S 173.16 W
	e	A	57 45.5	H = 20 37 53.9 h = normal MB=5.1 MS=4.8 D = 147.06 Az = 354.4 (NEIS) PKHKPV A 1.7s 45.5nm
27.	eP	A	23 16 14.5	Sicily 37.5 N 15.5 E
	LmH	B	21.1	H = 23 13 04 h = 0 km
	LmV	B	22.4	D = 13.38 Az = 349 (ISC) LmH B 16s 1.4 µm M = 4.1 LmV B 12 1.1 µm
28.	ePKHKP	A	05 06 39.5	South of Fiji Islands 23.28 S 176.15 W H = 04 46 49.2 h = normal MB = 5.1 D = 152.02 Az = 349.4 (NEIS)
28.	ePKP	A	13 19 06	New Hebrides Islands 19.20 S 169.82 E H = 12 59 29.7 h = 14 km MB = 5.5 D = 144.17 Az = 336.2 (NEIS)
28.	eP	A	13 43 02	Sicily 37.5 N 15.1 E H = 13 39 46 h = 0 km D = 13.42 Az = 350 (ISC)

Day	Phase		h m s	Remarks
28.	ePKIKP	ABC	14 19 17	Off E. Coast of N. Island, N.Z.
	ePKHKP	A	19 32.5	36.81 S 176.85 E
	ePKP2	ABC	20 06.5	H = 13 59 17.8 h = 15 km MB=5.8 MS=6.2
	ePP	ABC	23 56.5	D = 162.58 Az = 327.2 (NEIS)
	eSKSP	BC	34 18	PKP2V A 3.2s 1080.0nm
	ePPS	C	37 16	LmH B 19 4.8/um M = 6.2
	e	B	37 34	LmV B 19 6.2/um 6.5
	eSSS	BC	50 32	
	ePSPS2	BC	52 42	
	LmH	B	15 44.1	
	LmV	B	44.1	
28.	ePKP2	A	14 26 08	Off East Coast of North Island, N.Z.
				36.65 S 147.14 E
				H = 14 05 19 h = 15 km
				D = 162.57 Az = 328 (ISC)
				PKP2V A 1.6s 38.5nm
28.	ePKHKP	A	14 58 19	Off E.Coast of N. Island, N.Z.
				36.96 S 177.01 E
				H = 14 38 10.2 h = normal MB = 5.3
				D = 162.78 Az = 327.2 (NEIS)
28.	eP	A	15 11 39	Andreanof Islands, Aleutian Is.
				51.42 N 179.34 W
				H = 14 59 47.3 h = 56 km MB = 4.8
				D = 77.89 Az = 352.9 (NEIS)
				PV A 1.9s 30.3nm M = 5.0
28.	e	A	15 29 33	Upper Silesia, Poland
				Rockburst (WAR)
28.	eP	A	16 10 20	Hokkaido, Japan Region 42.17 N 143.10 E
				H = 15 58 27.8 h = 73 km MB = 4.7
				D = 78.33 Az = 330.9 (NEIS)

Day	Phase		h m s	Remarks
28.	ePKIKP	A	16 25 47	<u>Tonga Islands</u> 18.60 S 174.47 W H = 16 06 14.7 h = 98 km MB = 5.3 D = 147.66 Az = 352.7 (NEIS) PKIKPV A 1.7s 48.5nm
28.	e	A	17 25 15.5	<u>Southern Italy</u> 41.1 N 15.3 E H = 17 22 49 h = 178 km D = 9.90 Az = 346 (ISC)
28.	eP	A	19 31 12	<u>Fox Islands, Aleutian Is.</u> 53.01 N 166.66 W H = 19 19 21.9 h = normal MB = 5.0 D = 76.71 Az = 1.1 (NEIS) PV A 1.0s 27.6nm M = 5.2
28.	eP	A	20 28 16	<u>Costa Rica</u> 9.23 N 84.15 W H = 20 15 36.9 h = 59 km MB = 5.2 D = 86.58 Az = 39.4 (NEIS) PV A 1.3s 30.6nm M = 5.3
28.	eP	ABC	20 32 50	<u>Costa Rica</u> 9.34 N 84.06 W
	ePP	BC	36 10	H = 20 20 10.2 h = 46 km MB=5.8 MS=6.2
	iSKS	BC	43 24	D = 86.44 Az = 39.4 (NEIS)
	ePS	BC	44 28	PV A 2.0s 398.0nm M = 6.3
	eSS	BC	49 20	LmH B 19 7.3 μm 6.1
	eSSS	C	52 48	LmV B 17 8.2 μm 6.2
	LmH	B	21 06.7	
	LmV	B	16.4	
28.	eP	A	21 48 32	<u>Costa Rica</u> 9.47 N 83.93 W H = 21 35 54.7 h = 57 km MB = 4.7 D = 86.26 Az = 39.4 (NEIS)
28.	eP	A	22 25 07	<u>North of Svalbard</u> 81.59 N 1.98 W H = 22 18 48.8 h = normal MB = 4.5 D = 31.37 Az = 163.3 (NEIS)
28.	e	A	23 56 54	

Day	Phase		h m s	Remarks
1.	eP	A	03 09 39	Albania 40.59 N 19.93 E
				H = 03 06 52.1 h = normal
				D = 11.61 Az = 332.8 (NEIS)
1.	ePKP2	A	04 46 40	Off East Coast of North Island, N.Z.
				36.71 S 176.97 E
				H = 04 25 40.5 h = 17 km
				D = 162.54 Az = 328 (ISC)
1.	LmH	B	16 56.8	Komandorsky Islands Region
				54.38 N 169.57 E
				H = 16 04 02.0 h = 28 km MB=4.5 MS=5.4
				D = 73.73 Az = 345.6 (NEIS)
2.	ePKP2	A	05 06 47	Off E. Coast of N. Island, N.Z.
				36.8 S 177.0 E
				H = 04 45 58.4 h = 28 km MB = 5.4
				MS = 5.0 (NEIS)
				D = 162.6
				PKP2V A 1.4s 51.1nm
2.	ePKP2	A	12 29 29.5	Off East Coast of North Island, N.Z.
				36.63 S 177.18 E
				H = 12 08 37.6 h = 12 km
				D = 162.56 Az = 328 (ISC)
				PKP2V A 2s 42.7nm
3.	eP	ABC	05 03 09	Near East Coast of Honshu, Japan
	e	A	03 24	35.59 N 140.60 E
	ePP	BC	06 24	H = 04 50 48.9 h = 46 km MB=5.6 MS=5.6
	eS	BC	13 26	D = 83.09 Az = 330.1 (NEIS)
	LmH	B	42.9	PV A 2.0s 226.0nm M = 5.9
	LmV	B	45.3	PV B 6.4 1.74 μm 6.3
				LmH B 16.5 6.4 μm 6.1
				LmV B 13 4.2 μm 6.0
3.	eP	A	12 04 02	Kurile Islands 44.24 N 147.77 E
				H = 11 52 04.6 h = normal MB = 4.4
				D = 78.11 Az = 333.2 (NEIS)

Day	Phase	h m s	Remarks
3.	ePKP2 ABC	13 12 37	<u>Off East Coast of N. Island, N.Z.</u>
			36.7 S 177.1 E
			H = 12 51 44.9 h = 6 km MS = 5.4 (NEIS)
			D = 162.6
			PKP2V A 1.5s 37.7nm
3.	ePKP2 A	13 48 03.5	<u>Off East Coast of North Island, N.Z.</u>
			36.76 S 176.88 E
			H = 13 27 12.6 h = 10 km
			D = 162.55 Az = 327 (ISC)
3.	eiPKIKP ABC	14 42 14	<u>New Hebrides Islands</u> 20.06 S 169.71 E
	e B	42 50	H = 14 22 37.5 h = 17 km MB=6.1 MS=6.1
	e B	43 17	D = 144.91 Az = 335.6 (NEIS)
	ePPP A	48 43	PKIKPV A 1.4s 312.0nm
	LmH B	15 40.4	LmH B 19 4.0 µm M = 6.1
	LmV B	58.0	LmV B 18 2.9 µm 6.1
3.	eP A	16 28 20	<u>Svalbard Region</u> 76.55 N 13.10 E
			H = 16 22 48.0 h = normal MB = 4.2
			D = 26.01 Az = 182.2 (NEIS)
			PV A 1.4s 18.6nm M = 4.5
3.	ePKP A	18 27 29	<u>Fiji Islands Region</u> 16.74 S 176.78 E
			H = 18 07 54.9 h = 90 km MB = 4.9
			D = 144.14 Az = 343.8 (NEIS)
			PKPV A 1.7s 24.2nm
4.	eP A	01 10 55.5	<u>Azores Islands Region</u> 36.28 N 34.10 W
	LmV B	23.7	H = 01 03 59.7 h = normal MB=4.9 MS=4.8
	LmH B	24.2	D = 35.56 Az = 51.6 (NEIS)
			LmH B 16s 0.6 µm M = 4.5
			LmV B 18 0.8 µm 4.6
4.	ePn A	02 18 43.5	<u>Northern Italy</u> 46.02 N 10.54 E
	e A	19 10	H = 02 17 33.0 h = normal
			D = 4.68 Az = 8.4 (NEIS)

Day	Phase		h m s	Remarks
4.	eP	A	06 24 59	North of Severnaya Zemlya 85.37 N 92.95 E H = 06 17 33.4 h = normal MB = 4.4 D = 39.07 Az = 267.0 (NEIS)
4.	ePKP	A	11 35 02	New Hebrides Islands 20.05 S 169.52 E H = 11 15 29.8 h = 53 km MB = 5.1 D = 144.82 Az = 335.4 (NEIS) PKPV A 1.4s 23.2nm
4.	ePKHKP	ABC	12 57 34.5	Fiji Islands Region 18.84 S 177.66 W
	epPKP	ABC	59 08	H = 12 38 33.6 h = 383 km MB = 5.5
	e	A	13 02 27	D = 147.41 Az = 349.0 (NEIS)
	e	C	15 30	PKHKPV A 1.8s 118.0nm
6.	eP	A	00 15 37	Svalbard Region 77.05 N 12.72 E H = 00 10 00.3 h = normal MB = 3.9 D = 26.51 Az = 181.6 (NEIS)
6.	eP	ABC	01 52 53	Nicaragua 12.29 N 86.39 W
	epP	ABC	53 25	H = 01 40 26.4 h = 110 km MB = 5.8 D = 85.62 Az = 39.2 (NEIS) h = 128 km PV A 1.9s 90.9nm M = 5.4 pPV A 1.6 269.0nm
6.	eP	A	02 43 49	Ascension Island Region 5.64 S 11.36 W H = 02 33 47.5 h = normal MB = 5.2 D = 59.47 Az = 16.8 (NEIS) PV A 1.4s 23.2nm M = 5.1
6.	e	A	04 01 37	
6.	ePKIKP	A	04 38 41.5	New Hebrides Islands 18.69 S 169.15 E H = 04 19 38.0 h = 247 km MB = 4.8 D = 143.45 Az = 335.9 (NEIS) PKIKPV A 1.4s 32.6nm

Day	Phase		h m s	Remarks
6.	eP	A	06 43 41	Kurile Islands 44.16 N 148.02 E H = 06 31 42.8 h = normal MB = 5.0 D = 78.26 Az = 333.4 (NEIS)
6.	e(pPKP)	ABC	19 48 01	Banda Sea 6.60 S 128.98 E
	ePP	A	48 30	H = 19 29 08.1 h = 26 km MB=5.7 MS=6.3
	e	A	49 04	D = 112.26 Az = 322.3 (NEIS)
	e	BC	56 12	LmH B 19s 7.3 µm M = 6.3
	e	BC	57 44	LmV B 19.5 5.4 µm 6.2
	eSS	BC	20 04 24	
	LmH	B	36.0	
	LmV	B	42.4	
6.	eP	A	20 52 58	Near East Coast of Honshu, Japan
	e	A	53 17	40.28 N 142.22 E H = 20 40 55.1 h = 59 km MB = 5.0 D = 79.65 Az = 330.6 (NEIS)
7.	e	A	00 58 53	Banda Sea 6.43 S 129.20 E H = 00 39 55.7 h = 46 km MB=5.1 MS=4.7 D = 112.26 Az = 322.4 (NEIS)
7.	eP	A	03 54 57	Eastern Mediterranean Sea 33.92 N 25.54 E H = 03 50 29.0 h = normal MB = 4.3 D = 19.58 Az = 332.8 (NEIS)
7.	e	A	10 31 46	
7.	eP1	A	11 42 45	Iran-USSR Border Region
	eP2	A	42 48	37.60 N 55.83 E
	eS	B	48 12	H = 11 36 02.4 h = 21 km MB=5.1 MS=5.0
	LmH	B	12 02.4	D = 33.80 Az = 307.1 (NEIS)
	LmV	B	04.5	P2V A 1.4s 41.8nm M = 5.2
				LmH B 13 1.7 µm 5.0
				LmV B 14 1.4 µm 5.0

Day	Phase		h m s	Remarks
7.	ePKP2	A	19 02 38.5	Off East Coast of N. Island, N.Z. 36.72 S 176.93 E H = 18 41 48.0 h = 12 km D = 162.54 Az = 327 (ISC)
8.	eP	A	01 56 58.5	Afghanistan 33.10 N 69.38 E H = 01 48 41.7 h = normal MB = 5.0 D = 45.23 Az = 310.7 (NEIS)
8.	eP LmV LmH	A B B	02 38 07 46.5 46.6	Crete 34.78 N 24.72 E H = 02 33 53.1 h = 50 km MB = 4.5 D = 18.51 Az = 332.9 (NEIS) PV A 1.0s 11.8nm M = 4.0 LmV B 12 0.5/um 4.2
8.	ePg	A	08 00 20.5	Czechoslovakia 50.64 N 10.07 E Explosion H = 07 59.9 yield 4.5 t (PRU) D = 1.56 Az = 2.71 (ISC)
8.	eP epP LmH LmV	A A B B	09 37 29 37 43 10 19.3 19.3	Kyushu, Japan 30.92 N 131.63 E H = 09 25 06.1 h = 42 km MB=5.5 MS=5.4 D = 83.08 Az = 326.3 (NEIS) h = 50 km PV A 2.0s 42.7nm M = 5.2 LmH B 17 2.2/um 5.6 LmV B 17 2.5/um 5.7
8.	e(pP)	A	12 34 44.5	Crete 34.70 N 24.7 E H = 12 30 18 h = 44 km MB = 4.2 D = 18.56 Az = 333 (ISC)
9.	e(pP)	A	03 56 31	Crete 34.54 N 24.8 E H = 03 52 07 h = 38 km MB = 4.2 D = 18.77 Az = 333 (ISC)
9.	eP LmV	A B	04 16 23.5 24.8	Crete 34.63 N 25.05 E H = 04 12 07.7 h = 54 km MB = 4.6

Day	Phase		h m s	Remarks
cont.				
9.	LmH	B	04 25.0	D = 18.76 Az = 332.6 (NEIS)
				LmH B 11s 0.6 µm
				LmV B 11.5 0.8 µm
9.	iPKP	A	18 00 21	New Hebrides Islands 19.00 S 169.56 E
				H = 17 41 20.1 h = 286 km MB = 5.4
				D = 143.89 Az = 336.1 (NEIS)
				PKPV A 1.4s 93.0nm
9.	ePKP	A	18 27 07	New Hebrides Islands 19.02 S 169.35 E
				H = 18 07 59.6 h = 224 km
				D = 143.83 Az = 335.9 (NEIS)
9.	ePKIKP	ABC	20 33 31	Solomon Islands 7.48 S 156.18 E
	e	A	33 40	H = 20 14 28.3 h = 50 km MB=5.8 MS=6.5
	ePP	ABC	35 31.5	D = 127.89 Az = 332.1 (NEIS)
	e	BC	39 14	PKIKPV A 2.1s 148.3nm
	eSKS	BC	40 28	
	ePPS	BC	47 16	
9.	ePKIKP	A	20 37 05.5	Solomon Islands 7.34 S 156.19 E
				H = 20 18 66.3 h = normal MB=5.7 MS=6.6
				D = 127.77 Az = 332.2 (NEIS)
				PKIKPV A 2.0s 51.3nm
9.	ePKIKP	A	20 53 19	Solomon Islands 7.51 S 156.14 E
				H = 20 34 16.3 h = 48 km MB = 5.2
				D = 127.89 Az = 332.1 (NEIS)
9.	ePKHKP	A	24 16 57.0	Fiji Islands Region 20.27 S 178.46 W
				H = 23 58 13.1 h = 586 km MB = 4.5
				D = 148.66 Az = 347.6 (NEIS)
				PKHKPV A 1.4s 14.0nm
10.	eP	A	00 24 45	Andreanof Islands, Aleutian Is.
				50.53 N 175.11 W
				H = 00 12 40.4 h = 28 km MB = 4.7
				D = 79.04 Az = 355.6 (NEIS)

Day	Phase		h m s	Remarks
10.	ePKP	A	08 06 35	Solomon Islands 7.40 S 156.01 E
	LmH	B	09 06.5	H = 07 47 32.6 h = 54 km MB = 5.3
	LmV	B	06.7	D = 127.74 Az = 332.1 (NEIS)
				PKPV A 1.0s 21.6nm
				LmH B 20 0.5/um M = 5.2
				LmV B 18 0.6/um 5.3
10.	eP	ABC	16 30 12	Near Coast of Ecuador 0.40 N 80.05 W
	ePP	BC	34 00	H = 16 17 08.8 h = 43 km MB=5.1 MS=5.6
	eSKS	BC	40 48	D = 90.75 Az = 39.5 (NEIS)
	eS	BC	41 14	LmH B 19.5s 1.1/um M = 5.3
	LmV	B	17 16.3	LmV B 18 1.1/um 5.4
	LmH	B	16.4	
10.	ePP	A	21 54 09	Greece 40.75 N 21.16 E
				H = 21 51 02.5 h = 17 km MB = 4.1
				D = 11.92 Az = 329.3 (NEIS)
11.	ePKP	A	05 52 07	Scotia Sea 59.80 S 58.84 W
				H = 05 33 11.0 h = normal MB=5.4 MS=5.5
				D = 123.89 Az = 46.3 (NEIS)
11.	eP	AB	11 49 02.5	Kurile Islands 48.32 N 153.20 E
	e	B	49 38	H = 11 37 33.5 h = 169 km MB = 5.9
	e	B	52 28	D = 76.04 Az = 335.9 (NEIS)
	eS	B	58 28	PV A 2.4s 664.0nm M = 5.9
	eSS	B	12 03 28	SH B 11 2.3/um 6.1
	LmH	B	23.2	LmH B 18 3.6/um
11.	eP	A	20 28 50	Southern Iran 28.46 N 52.78 E
				H = 20 21 34.5 h = 44 km MB = 4.7
				D = 38.04 Az = 317.2 (NEIS)
12.	e	A	01 52 23	Iran 32.22 N 50.16 E
				H = 01 45 35.9 h = 34 km MB = 4.4
				D = 33.77 Az = 314.5 (NEIS)
				traces

Day	Phase		h m s	Remarks
12.	eP	A	06 59 22	Turkey-Iran Border Region
	LmH	B	07 11.6	38.41 N 44.02 E
	LmV	B	12.5	H = 06 53 52.1 h = 50 km MB = 4.5
				D = 25.93 Az = 308.7 (NEIS)
				LmH B 17s 0.8 µm M = 4.3
12.	eP	A	08 42 56.5	Kurile Islands 43.92 N 147.88 E
				H = 08 31 01.6 h = 57 km MB = 4.8
				D = 78.43 Az = 333.3 (NEIS)
12.	eP	A	10 27 21	Alaska Peninsula 54.31 N 162.41 W
				H = 10 15 41.2 h = 56 km MB = 4.0
				D = 75.30 Az = 3.9 (NEIS)
12.	eP	A	13 56 09.5	Kyushu, Japan 30.29 N 129.31 E
				H = 13 43 53.9 h = 80 km MB = 5.0
				D = 82.47 Az = 325.3 (NEIS)
12.	eP	A	18 20 30	Southwestern Ryukyu Islands
				23.56 N 125.40 E
				H = 18 07 56.0 h = 69 km MB = 4.8
				D = 85.93 Az = 324.3 (NEIS)
12.	eP	A	18 25 38	Dodecanese Islands 36.82 N 26.33 E
				H = 18 21 33.9 h = 39 km MB=4.7 MS=4.1
				D = 17.39 Az = 327.3 (NEIS)
				PV A 1.8s 33.8nm M = 4.2
13.	LmH	B	01 20.7	Talaud Islands 2.67 N 125.45 E
	LmV	B	24.7	H = 00 15 37.4 h = 51 km MB = 5.2 (ISC)
				D = 108.6
				LmH B 18s 0.6 µm
				LmV B 17 0.5 µm
13.	ePKIKP	A	08 28 47.5	Solomon Islands 7.39 S 156.10 E
				H = 08 09 46.2 h = 54 km MB = 5.4
				D = 127.77 Az = 332.1 (NEIS)
				PKIKPV A 1.4s 20.9nm

Day	Phase		h m s	Remarks
13.	ePKHKP	A	08 50 26	South of Fiji Islands 23.53 S 179.97 E H = 08 31 29.2 h = 494 km MB = 4.8 D = 151.46 Az = 344.4 (NEIS) PKHKPV A 1.2s 16.3nm
13.	e	A	10 01 26	Northern Italy 45.30 N 10.85 E H = 09 59 33.3 h = 0 km (ISC) D = 5.4
13.	e(Pg)	A	12 17 27	
13.	eP	A	17 25 02	Crete 34.65 N 24.83 E
	ePP	A	25 12.5	H = 17 20 45.5 h = 51 km MB = 4.6
	ePPP	A	25 25	D = 18.66 Az = 332.9 (NEIS)
	LmH	B	33.4	LmH B 14.5s 0.6/um M = 4.0
	LmV	B	33.4	LmV B 14 0.9/um 4.4
14.	ePKP	A	10 32 07.5	New Hebrides Islands 20.20 S 169.96 E H = 10 12 27.6 h = 1 km MB = 5.2 MS=5.0 D = 145.13 Az = 335.7 (NEIS) PKPV A 1.9s 136.0nm
14.	e(Pn)	A	21 00 15.5	Albania 41.85 N 19.38 E
	eSg	A	03 13	H = 20 57 34.4 h = normal MB = 4.3
	e	A	03 22	D = 10.30 Az = 331.2 (NEIS)
14.	ePKIKP	AB	21 18 20	New Hebrides Islands 13.86 S 166.77 E
	e	A	20 53	H = 20 58 54.8 h = 18 km MB=5.6 MS=5.8
	ePP	B	21 10	D = 138.13 Az = 336.4 (NEIS)
	e	B	40 54	PKIKPV A 1.6s 27.5nm
	LmH	B	22 27.3	LmH B 19 1.8/um M = 5.8
	LmV	B	27.8	LmV B 18 2.3/um 6.0
14.	ePKP	A	21 54 53	New Hebrides Islands Region 19.16 S 167.67 E H = 21 35 23.0 h = normal MB = 5.2 D = 143.28 Az = 334.4 (NEIS)

Day	Phase		h m s	Remarks
14.	ePKIKP'	A	23 12 50.5	Fiji Islands Region 19.16 S 177.83 W
	ePKHKP	A	12 54.5	H = 22 54 09.0 h = 542 km MB = 5.0
	ePKP2	A	12 58	D = 147.70 Az = 348.7 (NEIS)
				PKHKPV A 1.6s 33.0nm
				PKP2V A 1.2 28.4nm
15.	e	A	06 10 09	traces
15.	ePn	A	07 11 43	Poland 49.9 N 19.1 E
				H = 07 10 23 h = 0 km
				D = 4.82 Az = 281 (ISC)
15.	ePg	A	16 46 32.5	German Democratic Republic
				50.4 N 12.9 E
				H = 16 46 17
				D = 0.87 Az = 288 (ISC)
15.	ePcP	A	21 25 44	South of Honshu, Japan 30.50 N 141.73 E
				H = 21 12 39.6 h = 29 km MB = 4.8
				D = 87.97 Az = 330.8 (NEIS)
15.	eP	A	22 23 16.5	Kurile Islands Region 49.44 N 158.46 E
	LmH	B	45.7	H = 22 11 29.4 h = 35 km MB = 5.0
	LmV	B	46.5	D = 76.33 Az = 339.0 (NEIS)
				PV A 1.7s 24.2nm M = 4.9
				LmH B 16 0.5/um 4.9
				LmV B 16 0.7/um 5.1
16.	ePKP	A	01 22 56	Tonga Islands 15.26 S 173.63 W
				H = 01 03 24.5 h = 53 km MB = 4.7
				D = 144.45 Az = 354.2 (NEIS)
16.	ePKHKP	A	15 20 50	South of Fiji Islands 22.56 S 176.54 W
				H = 15 01 14.5 h = 160 km MB = 4.4
				D = 151.25 Az = 349.2 (NEIS)
17.	eP	A	01 31 43	Kurile Islands 43.79 N 147.66 E
				H = 01 19 51.3 h = 105 km MB = 4.7
				D = 78.47 Az = 333.2 (NEIS)

Day	Phase		h m s	Remarks
17.	eiP	A	04 09 48	Northern Sumatra 1.29 N 98.56 E H = 03 57 07.1 h = 61 km MB = 5.7 D = 87.06 Az = 320.5 (NEIS) PV A 1.8s 135.0nm M = 5.7
17.	ePKP	A	11 31 02	Loyalty Islands 20.94 S 168.77 E H = 11 11 18.9 h = 15 km MB = 4.8 D = 145.33 Az = 334.2 (NEIS)
17.	eP	A	13 36 10	Eastern Gulf of Aden 14.07 N 51.62 E H = 13 27 26.0 h = normal MB = 4.5 D = 48.74 Az = 327.0 (NEIS)
17.	eP	A	16 24 50.5	Ryukyu Islands 27.35 N 128.02 E H = 16 12 24.6 h = 74 km MB = 5.2 D = 84.23 Az = 325.0 (NEIS) PV A 1.2s 22.4nm M = 5.1
17.	ePKHKP	A	17 34 38	Tonga Islands Region 22.68 S 174.98 W H = 17 14 42.6 h = normal MB = 5.0 D = 151.61 Az = 351.2 (NEIS) PKHKPV A 1.4s 14.0nm
17.	e	A	24 02 15.5	Loyalty Islands Region 20.96 S 169.06 E H = 23 42 15.0 h = normal MB = 4.8 D = 145.46 Az = 334.5 (NEIS)
18.	ePKHKP	A	10 59 22	South of Fiji Islands 23.28 S 178.87 E H = 10 40 34.3 h = 592 km MB = 5.0 D = 150.93 Az = 343.2 (NEIS)
18.	ePKP LmH LmV	AB B B	11 15 45.5 12 23.0 23.2	Samoa Islands 14.93 S 172.83 W H = 10 56 12.4 h = 27 km MB=5.9 MS=6.0 D = 144.19 Az = 355.2 (NEIS) PKPV A 2.0s 158.0nm LmH B 19.5 4.2 μm M = 6.2 LmV B 18 4.0 μm 6.2

Day	Phase		h m s	Remarks
18.	e	A	11 26 30	
18.	eP	A	23 51 30.5	East China Sea 26.96 N 126.55 E H = 23 39 15.5 h = 140 km MB = 5.1 D = 83.79 Az = 324.5 (NEIS)
19.	ePKHKP	A	08 03 38.5	Fiji Islands Region 18.51 S 177.80 W H = 07 44 58.2 h = 570 km MB = 4.5 D = 147.07 Az = 349.0 (NEIS) PKHKPV A 1.2s 24.4nm
19.	eP	A	12 13 16	Near East Coast of Honshu, Japan 35.29 N 140.77 E H = 12 00 53.6 h = 58 km MB = 5.0 D = 83.41 Az = 330.2 (NEIS) traces
19.	eP	A	12 45 17.5	Taiwan Region 22.27 N 121.64 E H = 12 32 46.9 h = 64 km MB = 4.6 D = 84.92 Az = 323.1 (NEIS) traces
19.	eP	A	17 11 43	Crete 34.07 N 25.98 E H = 17 07 20.1 h = normal MB = 4.1 D = 19.62 Az = 331.9 (NEIS) traces
20.	ePKIKP	A	09 24 33	East New Guinea Region 5.52 S 147.48 E H = 09 05 58.6 h = 168 km MB = 5.2 D = 121.90 Az = 328.5 (NEIS) PKIKPV A 1.4s 18.6nm
20.	e(P)	A	17 05 51	Kurilen 44 N 147 E H = 16 53 42 (NORSAR) D = 77.9
20.	eP	A	18 55 38.5	North Atlantic Ocean 36.90 N 13.57 W H = 18 50 34.9 h = 0 km D = 22.65 Az = 45 (ISC) PV A 1.2s 10.2nm M = 4.3

Day	Phase		h m s	Remarks
20.	eP	A	23 42 22	Southern Greece 37.23 N 21.70 E H = 23 38 51.7 h = 111 km MB = 3.9 D = 15.23 Az = 334.9 (NEIS)
21.	eP LmV LmH	A B B	06 01 11.5 41.1 41.2	Near East Coast of Honshu, Japan 36.94 N 141.66 E H = 05 48 52.7 h = 43 km MB = 5.3 D = 82.34 Az = 330.5 (NEIS) PV A 1.5s 32.7nm M = 5.1 LmH B 14.8 0.9/um 5.3 LmV B 15 0.9/um 5.3
21.	eP	A	13 52 11.5	Central Mid-Atlantic Ridge 0.20 S 18.23 W H = 13 42 31.8 h = normal MB = 4.9 D = 56.67 Az = 22.3 (NEIS) PV A 1.7s 15.2nm M = 4.8
21.	eP	A	15 52 29	
21.	ePKP2	A	16 37 53.5	Tonga Region 19.0 S 172.4 W H = 16 18 14 h = 121 km D = 148.28 Az = 355 (ISC)
21.	ePKIKP e	A A	16 59 43 17 00 04	New Britain Region 4.71 S 152.68 E H = 16 40 52.4 h = 75 km MB = 4.9 D = 123.81 Az = 331.2 (NEIS)
22.	ePKP2	A	03 54 58.5	Kermadec Islands 29.2 S 177.3 W H = 03 34 37 D = 157.55 Az = 345 traces
22.	eP ePcP	A A	05 57 04.5 57 16	Unimak Island Region 53.70 N 163.43 W H = 05 45 19.0 h = normal MB = 4.9 D = 75.95 Az = 3.2 (NEIS) PV A 1.2s 28.4nm M = 5.1

Day	Phase		h m s	Remarks
22.	eP	A	07 15 51	Unimak Island Region 53.62 N 163.37 W
	ePcP	A	16 02	H = 07 04 06.2 h = normal MB=5.1 MS=4.6
				D = 76.02 Az = 3.3 (NEIS)
				PV A 1.8s 47.3nm M = 5.2
22.	eP	A	09 08 17	Taiwan 23.15 N 121.59 E
				H = 08 55 50.7 h = 63 km MB = 4.5
				D = 84.19 Az = 323.0 (NEIS)
22.	ePKHKP	A	14 32 43.5	Fiji Islands Region 21.06 S 178.86 W
				H = 14 14 01.4 h = 614 km MB = 4.7
				D = 149.35 Az = 346.9 (NEIS)
				PKHKPV A 1.6s 33.0nm
22.	eP	A	17 05 10	Greece - Albania Border Region
	ePP	A	05 20.5	40.68 N 20.49 E
	ePPPP	A	05 37.5	H = 17 02 20.8 h = normal MB = 4.7
				D = 11.73 Az = 331.1 (NEIS)
22.	eP	AB	18 22 20	USSR - Mongolia Border Region
	ePP	A	24 13.5	49.90 N 90.81 E
	LmH	B	41.7	H = 18 13 40.6 h = normal MB=5.5 MS=4.7
	LmV	B	43.2	D = 48.29 Az = 303.1 (NEIS)
				PV A 1.8s 67.6nm M = 5.4
				PPV A 1.5 55.3nm
				LmH B 12.2 1.9 µm 5.3
				LmV B 13 1.5 µm 5.2
22.	eP	A	19 15 38	Jan Mayen Island Region 70.74 N 14.72 W
				H = 19 10 27.6 h = 22 km MB = 5.0
				D = 23.54 Az = 135.0 (NEIS)
				PV A 2.0s 94.0nm M = 5.0
23.	eP	A	04 35 48	traces
23.	eP	A	07 03 14.5	Kurile Islands 46.86 N 153.79 E
				H = 06 51 20.4 h = normal MB = 4.8
				D = 77.54 Az = 336.4 (NEIS)

Day	Phase		h m s	Remarks
23.	eP	A	07 11 03.5	North Atlantic Ocean 54.02 N 36.09 W H = 07 05 06.9 h = normal MB=4.6 MS=4.6 D = 28.91 Az = 77.0 (NEIS) PV A 2.0s 25.6nm M = 4.6
23.	e(P)	A	07 15 03	North Atlantic Ocean 53.88 N 35.33 W H = 07 09 02.8 h = normal MB=5.0 MS=4.9 D = 28.50 Az = 77.1 (NEIS) (P)V A 2.2s 32.7nm (M = 4.6)
23.	eP	A	07 25 09	North Atlantic Ocean 53.84 N 35.42 W
	LmH	B	35.9	H = 07 19 14.2 h = normal MB=5.1 MS=5.1
	LmV	B	36.5	D = 28.57 Az = 77.0 (NEIS) PV A 2.2s 65.5nm M = 4.9 LmH B 16.5 1.4/um 4.7 LmV B 15.5 1.6/um 4.8
23.	e	A	10 17 20.5	Probably North Western Kashmir 35.6 N 74.6 E H = 10 08 51 h = 228 km (ISC) D = 46.9
23.	+iPKIKP	AB	14 47 23	South of Fiji Islands 23.93 S 179.78 E
	epPKP	AB	49 30	H = 14 28 35.4 h = 535 km MB = 6.1
	ePP	B	53 14	D = 151.79 Az = 344.0 (NEIS)
	esPKS	B	57 16	PKIKPV A 2.2s 895.0nm
	e	B	15 00 47	PKIKPV B 7.4 6.2/um
	ePPS	B	03 24	
	eSS	B	09 42	
23.	ePKIKP	A	15 12 56	South of Fiji Islands 23.80 S 178.76 E
	ePKHKP	A	13 02	H = 14 54 11.3 h = 569 km MB = 5.6
	ePKP2	A	13 13.5	D = 151.40 Az = 342.8 (NEIS)
	epPKIKP	A	15 05.5	h = 546 km
	epPKHKP	A	15 12	PKIKPV A 2.2s 81.8nm PKHKPV A 1.2 110.0nm PKP2V A 1.2 81.3nm pPKHKPV A 1.6 88.0nm

Day	Phase		h m s	Remarks
23.	ePKP	A	15 19 56	South of Fiji 23.6 S 177.1 W H = 15 00 19 h = 0 km MB = 5.1 D = 152.20 Az = 348 (ISC) PKPV A 1.4s 14.0nm
23.	ePKHKP	A	15 31 17	South of Fiji 26.3 S 179.0 W H = 15 11 56 h = 320 km D = 154.34 Az = 344 (ISC) PKPV A 1.4s 14.0nm
23.	ePKP	A	16 07 23	South of Fiji 24.1 S 179.7 E H = 15 48 49 h = 557 km D = 151.90 Az = 344 (ISC) PKPV A 2.2s 32.7nm
23.	ePg	A	18 09 40.5	Poland 50.3 N 19.5 E H = 18 08 07 h = 0 km D = 5.05 Az = 277 (ISC) PV A 1.4s 14.0nm
23.	ePKHKP	A	18 32 31	South of Fiji Islands 23.86 S 179.65 E H = 18 13 35.7 h = 531 km MB = 4.6 D = 151.69 Az = 343.8 (NEIS)
23.	eP	A	19 21 49	Off Coast of Oregon 42.68 N 126.09 W H = 19 09 40.9 h = normal MB = 4.5 D = 80.01 Az = 25.8 (NEIS)
23.	ePKHKP	A	20 30 32	South of Fiji Islands 23.91 S 179.75 E H = 20 11 36.4 h = 532 km MB = 4.7 D = 151.77 Az = 343.9 (NEIS)
23.	ePKHKP LmV LmH	A B B	20 45 34.5 21 48.9 50.2	New Hebrides Islands Region 21.93 S 173.71 E H = 20 25 51.8 h = normal MB=5.7 MS=6.1 D = 148.09 Az = 338.3 (NEIS) PKHKPV A 1.7s 97.0nm LmH B 22 4.2/um M = 6.1 LmV B 23 4.8/um 6.2

Day	Phase		h m s	Remarks
23.	ePKIKP	A	21 10 55	South of Fiji Islands 23.95 S 179.51 E
	ePKHKP	A	11 02	H = 20 52 06.8 h = 521 km MB = 5.0
	ePKP2	A	11 13.5	D = 151.74 Az = 343.6 (NEIS)
	epPKP	A	13 13	PKHKPV A 1.1s 20.2nm
23.	ePKHKP	A	22 05 57	Tonga Islands 20.72 S 175.17 W
				H = 21 46 12.0 h = normal MB=5.3 MS=5.4
				D = 149.66 Az = 351.4 (NEIS)
				PKHKPV A 1.6s 24.7nm
23.	eP	A	22 27 03	Tsinghai Province, China 37.00 N 95.74 E
				H = 22 17 03.9 h = normal MB = 4.8
				D = 59.06 Az = 312.4 (NEIS)
24.	ePKHKP	A	00 31 25.5	South of Fiji Islands 21.15 S 175.67 E
				H = 00 12 46.0 h = 600 km MB = 4.6
				D = 148.00 Az = 340.7 (NEIS)
				PKHKPV A 2.0s 132.0nm
24.	eP	A	04 35 04.5	South of Mariana Islands
	ePKKP	A	50 49.5	12.55 N 144.26 E
	LmH	B	05 17.9	H = 04 21 05.8 h = 79 km MB = 5.9
	LmV	B	24.5	D = 104.75 Az = 331.0 (NEIS)
				PV A 1.7s 24.2nm M = 5.9
				PKKPV A 1.6 27.5nm
				LmH B 20 3.1 μm
				LmV B 14 1.2 μm
24.	eP	AB	14 26 04.5	Nepal 27.73 N 86.11 E
	eS	B	34 11	H = 14 16 03.1 h = normal MB=5.7 MS=5.7
	eScS	B	35 54	D = 59.49 Az = 314.6 (NEIS)
	LmH	B	54.6	PV A 0.8s 65.4nm M = 5.8
	LmV	B	54.6	LmH B 15.5 10.2 μm 6.1
				LmV B 15.5 13.0 μm 6.2
24.	eP	A	16 27 41.5	Nepal 27.59 N 86.02 E
				H = 16 17 39.6 h = normal MB = 4.8
				D = 59.52 Az = 314.6 (NEIS)

Day	Phase		h m s	Remarks
24.	ePKP	A	19 59(44)	Tonga Islands 17.75 S 174.19 W H = 19 40 14.8 h = 165 km. MB = 4.8 D = 146.86 Az = 353.2 (NEIS) traces
25.	ePKP2	A	04 27 59	Fiji Islands Region 19.35 S 176.23 W H = 04 08 11.0 h = normal MB=5.0 MS=5.2 D = 148.16 Az = 350.5 (NEIS)
25.	eP LmH LmV	A B B	07 50 10.5 08 36.9 38.6	Near Coast of Michoacan, Mexico 18.18 N 103.33 W H = 07 37 12.3 h = 87 km MB = 4.7 D = 90.92 Az = 35.3 (NEIS) LmH B 14s 1.7/um LmV B 15 2.2/um
25.	ePKIKP	A	12 03 50.5	Solomon Islands 6.70 S 155.14 E H = 11 44 52.3 h = 58 km MB = 5.1 D = 126.72 Az = 331.8 (NEIS)
25.	e	A	19 44 11.5	South Atlantic Ridge 32.48 S 14.18 W H = 19 31 21.3 h = normal MB = 4.7 D = 85.85 Az = 16.1 (NEIS)
25.	eP	A	21 26 07	Burma 17.41 N 94.00 E H = 21 14 44.2 h = normal MB = 4.9 D = 71.95 Az = 318.4 (NEIS)
26.	ePKIKP	A	00 27 38	South of Fiji 24.1 S 179.41 W H = 00 07 47.2 h = 0 km MB = 4.9 D = 152.11 Az = 345 (ISC)
26.	ePKHKP	A	09 36 45	Tonga Islands 19.81 S 173.04 W H = 09 16 58.5 h = normal MB = 4.9 D = 149.01 Az = 354.2 (NEIS) traces
26.	e	A	11 04 40.5	

Day	Phase		h m s	Remarks
27.	ePKIKP	A	03 27 12.5	Kermadec Islands Region 27.27 S 175.34 W H = 03 07 22.3 h = 44 km MB=5.7 MS=5.0 D = 156.06 Az = 349.1 (NEIS) PKIKPV A 1.3s 17.5nm
27.	ePKP	A	16 33 35.5	Tonga Islands 15.68 S 173.67 W H = 16 14 15.8 h = 136 km MB = 4.8 D = 144.86 Az = 354.1 (NEIS) PKPV A 1.4s 14.0nm
27.	eP	A	16 40 50	Andreanof Islands, Aleutian Is. 50.11 N 179.66 W H = 16 28 47.3 h = 37 km MB=5.6 MS=4.8 D = 79.16 Az = 352.7 (NEIS) PV A 1.4s 83.6nm M = 5.5
27.	ePKHKP	A	17 58 55	South of Tonga 26.9 S 174.6 W H = 17 38 53 D = 155.78 Az = 350 (ISC)
28.	ePKIKP ePKHKP	A A	02 18 53 18 58	Fiji Islands Region 20.38 S 176.96 W H = 01 59 42.4 h = 287 km MB = 5.0 D = 149.05 Az = 349.4 (NEIS)
28.	ePKP2	A	05 17 55.5	Norfolk Island Region 27.27 S 170.81 E H = 04 57 50.3 h = normal D = 151.79 Az = 331.4 (NEIS)
28.	eP	A	17 38 51	Queen Elizabeth Islands 76.48 N 106.78 W H = 17 30 18.6 h = normal MB = 4.3 D = 47.29 Az = 49.7 (NEIS)
28.	eP LmH LmV	A B B	21 35 49.5 40.8 42.1	Sicily 37.15 N 14.99 E H = 21 32 35.3 h = normal MB = 4.6 D = 13.71 Az = 350.9 (NEIS) PV A 1.3s 10.9nm M = 4.5 LmH B 18 0.9 µm 3.9 LmV B 12 0.4 µm

Day	Phase		h m s	Remarks
29.	ePg	A	09 20 28.5	Czechoslovakia 35.59 N 13.83 E
	eSg	A	20 48.5	H = 09 20.0
				Explosion yield 8.3 t (PRU)
				D = 1.41 Az = 273 (ISC)
29.	ePKHKP	A	17 05 36.5	Tonga Islands 21.89 S 175.60 W
				H = 16 45 46.6 h = 68 km MB = 5.0
				D = 150.74 Az = 350.6 (NEIS)
29.	eP	A	20 53 43	West Pakistan 26.62 N 66.24 E
				H = 20 45 05.3 h = normal MB = 4.7
				D = 47.65 Az = 315.4 (NEIS)
				PV A 0.9s 15.6nm M = 5.0
29.	iP	AB	22 01 53.5	Kodiak Island Region 57.59 N 153.92 W
	epP	AB	02 04	H = 21 50 35.3 h = 44 km MB=5.7 MS=5.2
	eS	AC	11 10	D = 71.48 Az = 9.7 (NEIS)
	LmH	C	32.3	h = 38 km
				PV A 2.0s 384.0nm M = 6.0
				PV B 3.2 0.7 µm 6.0
				LmH C 24 0.9 µm 6.0
30.	e	A	00 40 29	Western Caucasus 41.5 N 44.3 E
				H = 00 35 05 h = 33 km MB = 4.0
				D = 24.28 Az = 303 (ISC)
				traces
30.	ePKHKP	A	02 10 52	South of Fiji Islands 23.46 S 179.99 E
				H = 01 52 01.0 h = 572 km MB = 4.7
				D = 151.39 Az = 344.5 (NEIS)
30.	eP	A	06 34 33	Pakistan 26.70 N 66.51 E
				H = 06 25 55.4
				D = 47.77 Az = 315 (ISC)
30.	eP	A	08 35 18	Central Mid-Atlantic Ridge
				3.90 N 32.17 W
				H = 08 25 17.4 h = normal MB = 5.0
				D = 59.29 Az = 30.8 (NEIS)

Day	Phase		h m s	Remarks
30.	eP	A	08 46 10.5	Kurile Islands 43.47 N 146.13 E H = 08 34 14.5 h = 55 km MB = 4.7 D = 78.25 Az = 332.4 (NEIS) traces
30.	eP	A	18 46 28	Iceland Region 63.88 N 23.10 W
	e	A	46 38	H = 18 41 26.3 h = normal MB = 4.6
	LmV	B	57.9	D = 22.61 Az = 109.4 (NEIS)
	LmH	B	58.2	LmH B 12s 0.6 µm M = 4.3 LmV B 12.5 0.9 µm 4.5
30.	eP	A	19 06 38	Iceland Region 68.3 N 23.2 W H = 19 01 53 h = 0 km D = 22.62 Az = 109 (ISC)
30.	eP	A	19 15 00	Iceland Region 63.78 N 23.37 W
	LmV	B	26.5	H = 19 09 59.6 h = normal MB = 4.4
	LmH	B	26.9	D = 22.69 Az = 108.9 (NEIS) LmH B 10s 0.5 µm M = 4.2 LmV B 12 0.7 µm 4.4
30.	ePKIKP	A	19 25 50.5	Fiji Islands Region 16.39 S 176.80 E H = 19 06 10.5 h = normal D = 143.81 Az = 344.0 (NEIS)
30.	eP	A	19 38 04	Off East Coast of Honshu, Japan 40.51 N 143.58 E H = 19 25 56.5 h = 38 km MB = 4.7 D = 79.96 Az = 331.3 (NEIS)
30.	eP	A	20 21 37	Iceland Region 63.38 N 23.63 W
	LmH	B	33.0	H = 20 16 36.7 h = normal MB = 4.3
	LmV	B	33.2	D = 22.68 Az = 107.6 (NEIS) PV A 1.3s 17.5nm M = 4.4 LmH B 11.5 0.5 µm 4.2 LmV B 12 0.5 µm 4.3

Day	Phase		h m s	Remarks
31.	ePKIKP	A	05 08 02.5	Southern Pacific Ocean 35.98 S 103.18 W H = 04 48 50.2 h = normal MB=5.2 MS=5.2 D = 131.89 Az = 50.9 (NEIS)
31.	ePKP	A	07 06 28	Near N. Coast of West New Guinea 2.19 S 139.08 E H = 06 47 45.0 h = normal MB=5.6 MS=5.6 D = 114.60 Az = 326.2 (NEIS) LmH B 19s 2.0/um M = 5.8 LmV B 19 2.8/um 5.9
	LmV	B	59.2	
	LmH	B	59.3	
31.	eP	A	21 21 14	North Atlantic Ocean 17.04 N 26.44 W H = 21 12 57.8 h = normal MB = 4.9 D = 45.24 Az = 33.6 (NEIS) PV A 1.7s 30.3nm M = 4.9
31.	eP	A	21 26 27	Near Coast of Nicaragua 11.80 N 87.89 W H = 21 13 45.8 h = 49 km MB=5.0 MS=4.7 D = 86.92 Az = 39.0 (NEIS) PV A 1.4s 14.0nm M = 5.0 LmH C 18 0.6/um 4.9
	LmH	C	55.3	

Day	Phase		h m s	Remarks
1.	eP	A	00 26 34	Mediterranean Sea 35.56 N 22.44 E
	LmH	B	34.4	H = 00 22 39.8 h = 67 km MB = 4.4
	LmV	B	34.4	D = 16.99 Az = 335.9 (NEIS)
				PV A 2.0s 34.2nm M = 4.2
				LmH B 14 1.1 /um 4.3
				LmV B 16 1.1 /um 4.3
1.	eP	A	15 46 42	Hokkaido, Japan Region 43.19 N 145.86 E
				H = 15 34 48.9 h = 84 km MB = 4.5
				D = 78.40 Az = 332.3 (NEIS)
1.	ePKIKP	A	15 50 35.5	Tonga Islands 18.56 S 175.39 W
	e	A	50 51	H = 15 31 10.7 h = 180 km MB = 4.9
				D = 147.50 Az = 351.7 (NEIS)
				traces
1.	eP	A	22 03 39	South of Honshu, Japan 30.98 N 141.99 E
	e	A	03 50.5	H = 21 50 49.5 h = 16 km MB=5.3 MS=5.3
	ePP	A	07 02	D = 87.66 Az = 331.0 (NEIS)
	eS	BC	14 18	PV A 1.6s 35.7nm M = 5.4
	LmH	B	43.8	PPV A 1.3 21.8nm 5.5
	LmV	B	50.7	LmH B 15 1.5 /um 5.5
				LmV B 14 1.7 /um 5.6
2.	eP	A	03 36 26.5	Hokkaido, Japan Region 41.25 N 141.64 E
				H = 03 24 27.4 h = 41 km MB=5.2 MS=5.4
				D = 78.60 Az = 330.2 (NEIS)
				PV A 1.1s 14.1nm M = 4.9
2.	ePKIKP	A	04 21 33	Solomon Islands 6.95 S 155.29 E
	e	A	21 45	H = 04 02 33.6 h = 47 km MB=5.4 MS=5.7
				D = 127.00 Az = 331.8 (NEIS)
				PKIKPV A 1.2s 20.3nm
2.	ePKHKP	A	06 22 18	South of Fiji Islands 23.81 S 179.72 E
				H = 06 03 21.3 h = 504 km MB = 4.3
				D = 151.66 Az = 343.9 (NEIS)

Day	Phase		h m s	Remarks
2.	eP	A	08 11 04	Kurile Islands 46.60 N 153.84 E H = 07 59 08.0 h = normal MB = 4.3 D = 77.79 Az = 336.5 (NEIS) traces
2.	epP	A	20 25 24	Near Coast of Guatemala 13.56 N 90.26 W H = 20 12 21.1 h = 91 km MB = 4.6 D = 87.00 Az = 38.6 (NEIS) h = 100 km
3.	e	A	04 39 51	Hokkaido Region 42.55 N 141.38 E
	e	A	45 21	H = 04 27 13.3 h = 123 km MB = 4.7 (ISC)
	LmH	B	05 21.4	D = 77.2
	LmV	B	23.9	LmH B 21s 1.2/um LmV B 24 1.4/um
3.	LmH	C	23 41.0	South of the Marianas 13.07 N 143.36 E H = 04 22 31 h = 204 km (ISC) D = 104.0 LmH C 22s 0.35/um
4.	eP	A	04 28 02	Tadzhik SSR 39.20 N 71.73 E
	e	A	29 51	H = 04 20 01.8 h = 21 km MB = 5.1 D = 42.95 Az = 305.9 (NEIS) PV A 1.0s 11.8nm M = 4.6
4.	LmH	B	06 28.5	LmV B 20s 0.5/um
	LmV	B	28.5	
4.	eP	A	07 49 09	Honshu, Japan 37.72 N 140.77 E H = 07 37 02.6 h = 97 km MB = 5.3 D = 81.31 Az = 330.0 (NEIS) PV A 1.4s 32.6nm M = 5.0
4.	eP	A	15 00 57.5	South Atlantic Ridge 12.65 S 14.07 W
	e	A	01 04.5	H = 14 50 06.1 h = normal MB = 5.2 D = 66.91 Az = 17.5 (NEIS) PV A 2.0s 47.0nm M = 5.2

Day	Phase		h m s	Remarks
4.	ePKHKP	A	18 38 30.5	Fiji Islands Region 20.89 S 178.67 W
				H = 18 19 47.4 h = 584 km MB = 4.4
				D = 149.21 Az = 347.2 (NEIS)
5.	eP	A	03 57 31.5	Burma 21.33 N 93.54 E
	ePcP	A	57 53	H = 03 46 30.1 h = 49 km MB = 5.0
				D = 68.76 Az = 317.4 (NEIS)
				PV A 1.0s 11.8nm M = 4.8
5.	eP	A	05 04 31.5	North Atlantic Ridge 28.65 N 43.61 W
	LmH	B	21.5	H = 04 56 05.1 h = 33 km MB = 4.9
	LmV	B	21.7	D = 46.61 Az = 46 (ISC)
				LmH B 17s 0.45 µm M = 4.5
				LmV B 18 0.55 µm 4.7
5.	eP	A	08 04 03.5	Taiwan Region 22.15 N 121.30 E
				H = 07 51 33.4 h = 58 km MB = 4.7
				D = 84.82 Az = 323.0 (NEIS)
				PV A 1.3s 15.3nm M = 4.9
5.	ePKP	A	12 02 22	Tonga Islands 17.43 S 174.27 W
	e	A	02 55	H = 11 42 48.1 h = 79 km MB = 4.8
				D = 146.53 Az = 353.2 (NEIS)
				PKPV traces
6.	+eiP	A	02 05 24	Alaska Peninsula 55.10 N 160.44 W
	epP	BC	05 40	H = 01 53 47.3 h = 27 km MB=5.7 MS=5.1
	eS	C	14 52	D = 74.41 Az = 5.2 (NEIS)
	LmH	C	32.5	h = 58 km
	LmV	C	37.3	PV A 0.9s 347.0nm M = 6.4
				LmH C 30 0.9 µm 4.9
				LmV C 22 0.35 µm 4.6
6.	ePKIKP	A	02 31 11	New Hebrides Islands 14.61 S 166.84 E
	e(PKS)	A	34 53	H = 02 11 40.3 h = 8 km MB=5.3 MS=5.2
				D = 138.84 Az = 336.1 (NEIS)
6.	eP	A	02 38 56.5	Alaska Peninsula 55.34 N 160.60 W
				H = 02 27 21.8 h = normal MB = 4.3
				D = 74.19 Az = 5.1 (NEIS)

Day	Phase		h m s	Remarks
6.	eP	A	02 47 02.5	North Atlantic Ridge 49.91 N 29.01 W
				H = 02 41 32.2 h = normal MB = 4.7
				D = 25.75 Az = 72.6 (NEIS)
				PV A 1.3s 19.7nm M = 4.5
6.	ePKIKP	A	03 02 53	New Hebrides Islands 14.61 S 166.69 E
	e(PKS)	A	06 33	H = 02 43 25.2 h = 16 km MB=5.4 MS=5.2
	LmH	C	04 06.0	D = 138.78 Az = 336.0 (NEIS)
	LmV	C	06.0	PKIKPV A 1.9s 37.9nm
				LmH C 20 0.6/um M = 5.3
				LmV C 20 0.3/um 5.0
6.	ePKIKP	A	04 03 12	New Hebrides Islands 14.59 S 166.81 E
				H = 03 43 46.2 h = 19 km MB = 5.2 (NEIS)
				D = 138.81
6.	+iP	AB	04 07 37	Alaska Peninsula 55.12 N 160.44 W
	epP	A	07 48	H = 03 56 01.8 h = 40 km MB=6.0 MS=5.3
	esP	AB	07 53.5	D = 74.40 Az = 5.2 (NEIS)
	eS	C	17 00	h = 44 km
	LmH	B	43.4	PV A 0.9s 642.0nm M = 6.6
	LmV	B	47.6	LmH B 20.5 1.0/um 5.1
				LmV B 18 0.6/um 5.0
6.	eP	A	05 23 44.5	Kodiak Island Region 57.80 N 153.50 W
	epP	A	23 53.5	H = 05 12 26.4 h = 53 km MB = 4.6
				D = 71.23 Az = 9.9 (NEIS)
				h = 46 km
6.	ePKHKP	A	06 19 56.5	Fiji Islands Region 20.36 S 178.18 W
				H = 06 01 11.4 h = 579 km MB = 4.7
				D = 148.80 Az = 347.9 (NEIS)
				PKHKPV A 1.6s 41.2nm
6.	ePKIKP	A	08 10 49	New Hebrides Islands 14.66 S 166.72 E
	LmH	C	09 05.2	H = 07 51 21.2 h = 18 km MB=5.3 MS=5.2
	LmV	C	05.5	D = 138.84 Az = 336.0 (NEIS)
				PKIKPV A 1.3s 13.1nm
				LmH C 21 0.4/um M = 5.1
				LmV C 21 0.5/um 5.2

Day	Phase		h m s	Remarks
6.	ePKP2	A	12 37 09	Auckland Islands Region
	e	A	37 17	49.14 S 164.67 E
				H = 12 16 18.5 h = 12 km
				D = 162.60 Az = 285 (ISC)
				PKP2V A 2.0s 34.2nm
6.	-eP	A	20 27 43.5	Tadzhik SSR 37.05 N 72.49 E
	LmH	B	48.5	H = 20 19 36.9 h = 84 km MB = 5.2
	LmV	B	48.6	D = 44.72 Az = 307.8 (NEIS)
				PV A 1.8s 60.8nm M = 5.1
				LmH B 14.5 0.5/um
				LmV B 11 0.8/um
6.	eP	A	22 19 09	Kurile Islands 43.43 N 146.29 E
				H = 22 07 13.5 h = 53 km MB = 4.9
				D = 78.34 Az = 332.5 (NEIS)
				PV A 1.3s 13.1nm M = 4.7
7.	e(P)	A	01 00 37.5	Algeria 36 1/2 N 4.1 E
				H = 00 57 00 (BCIS)
				D = 15.23
7.	ePKP2	A	03 35 36.5	Kermadec Islands Region
				27.39 S 177.69 W
				H = 03 15 32.9 h = 170 km MB = 5.2
				D = 155.72 Az = 345.5 (NEIS)
				traces
7.	eP	A	06 13 52.5	Kurile Islands 46.79 N 153.64 E
				H = 06 01 56.6 h = normal MB = 4.5
				D = 77.56 Az = 336.3 (NEIS)
7.	eP	ABC	14 27 02.5	Crete 34.75 N 24.70 E
	eS	B	30 38	H = 14 22 47.1 h = 29 km MB=4.7 MS=5.0
	LmH	B	35.5	D = 18.52 Az = 333.0 (NEIS)
	LmV	B	35.6	PV B 6s 0.6/um M = 4.9
				LmH B 13.5 5.1/um 5.0
				LmV B 12 6.1/um 5.3

Day	Phase		h m s	Remarks
7.	eP	A	16 16 08	**West Pakistan** 32.04 N 69.73 E H = 16 07 45.8 h = 44 km MB = 4.8 D = 46.15 Az = 311.4 (NEIS)
8.	eP	A	00 12 29.5	**Albania** 40.57 N 19.83 E H = 00 09 42.2 h = 19 km MB = 3.2 D = 11.60 Az = 333.1 (NEIS)
8.	eP	A	07 30 39	**Costa Rica** 9.50 N 84.32 W
	e	A	30 48	H = 07 17 58.6 h = normal MB = 4.5 D = 86.47 Az = 39.4 (NEIS) PV traces
8.	ePn	A	08 30 23.5	**Czechoslovakia** 50.69 N 14.66 E
	eSg	A	30 51	Explosion yield 14.6 t (PRU) D = 1.94
8.	eP	A	17 59 03.5	**Southern Sumatra** 2.26 S 99.53 E
	e	A	59 11.5	H = 17 46 01.9 h = normal MB = 4.7 D = 90.40 Az = 320.5 (NEIS) PV A traces
8.	e	A	22 18 46	**Southern Sumatra** 2.13 S 99.71 E H = 22 05 19.6 h = normal MB = 4.3 D = 90.42 Az = 320.5 (NEIS)
9.	eP	A	00 34 10	**Greece-Albania Border Region** 39.44 N 20.24 E H = 00 31 09.0 h = 44 km MB = 3.1 D = 12.75 Az = 334.4 (NEIS) PV A traces
9.	e	A	04 26 28.5	
9.	eiP	AB	13 23 01.5	**Kurile Islands** 45.49 N 148.29 E H = 13 11 21.6 h = 139 km MB = 5.5 D = 77.16 Az = 333.4 (NEIS) PV A 1.2s 199.2nm M = 5.7

Day	Phase		h m s	Remarks
9.	eP	A	21 03 05.5	<u>Rat Islands, Aleutian Is.</u> 51.36 N 177.82 E H = 20 51 11.0 h = 37 km MB=4.7 MS=4.7 D = 77.70 Az = 351.1 (NEIS)
10.	eP	A	01 40 11	<u>Fox Islands, Aleutian Is.</u> 52.52 N 168.74 W H = 01 28 15.9 h = 15 km MB = 4.6 D = 77.22 Az = 359.8 (NEIS) PV A 1.1s 16.1nm M = 5.0
10.	eP	AB	22 55 36.5	<u>Guatemala</u> 14.53 N 91.63 W
	esP	A	56 14.5	H = 22 43 00.6 h = 108 km MB = 5.4
	eSKS	BC	23 05 56	D = 87.07 Az = 38.4 (NEIS)
	eSS	BC	11 52	h = 100 km
	LmH	B	34.8	PV A 1.4s 46.5nm M = 5.3
	LmV	B	38.9	LmH B 20 0.9/um
				LmV B 17 0.6/um
10.	eP	A	23 12 19	<u>Taiwan Region</u> 22.77 N 121.30 E
	LmH	B	55.1	H = 22 59 51.1 h = 53 km MB = 4.9
	LmV	B	55.1	D = 84.33 Az = 323.0 (NEIS)
				LmH B 17s 1.2/um
				LmV B 15 1.8/um
11.	eP	A	21 49 48.5	<u>Hokkaido, Japan Region</u> 42.40 N 144.43 E H = 21 37 53.0 h = 75 km MB = 5.3 D = 78.61 Az = 331.5 (NEIS) PV A 1.5s 32.7nm M = 5.0
12.	ePn	A	07 45 33	<u>Czechoslovakia</u> 51.00 N 14.41 E
	eiSg	A	45 52	H = 07 45.0 yield 21 t (KHC)
				D = 1.81
12.	ePKP	A	12 35 19	<u>New Hebrides Islands</u> 18.64 S 169.22 E H = 12 16 15.3 h = 244 km MB = 5.1 D = 143.43 Az = 336.0 (NEIS) PKPV A 1.3s 28.4nm

Day	Phase		h m s	Remarks
12.	eP	A	17 58 54	Philippine Sea 14.26 N 134.32 E
	LmH	C	18 45.0	H = 17 45 18.1 h = normal MB=5.5 MS=4.9
				D = 98.32 Az = 327.2 (NEIS)
				PV A 1.1s 16.1nm M = 5.5
				LmH C 21 0.4/um 4.9
13.	eP	A	05 11 04.5	Lake Baikal Region 55.74 N 111.04 E
				H = 05 01 35.6 h = normal MB = 4.9
				D = 54.76 Az = 309.7 (NEIS)
				PV A 1.2s 14.2nm M = 4.9
13.	ePn	A	15 33 35	Poland 50.32 N 18.83 E
				H = 15 31 20.6 M = 3.0 (WAR)
				D = 4.62 Az = 277 (ISC)
13.	eP	A	20 20 30.5	Near East Coast of Kamchatka
	e	A	20 44	55.29 N 161.84 E
				H = 20 09 10.6 h = normal MB=4.8 MS=4.2
				D = 71.55 Az = 340.5 (NEIS)
				PV A 1.4s 20.9nm M = 5.0
14.	eP	A	00 24 09.5	Near East Coast of Kamchatka
				55.8 N 161 E
				H = 00 12 44 h = 0 km MB = 4.4 (NORSAR)
				D = 72.20 Az = 341 (ISC)
14.	ePKP	A	01 30 56	Loyalty Islands 20.94 S 168.72 E
	eX	A	31 01	H = 01 11 15.1 h = 4 km MB = 4.4
	eY	A	31 07	D = 145.30 Az = 334.2 (NEIS)
				PKPV A 2.0s 29.9nm
				XV A 1.3 26.2nm
				YV A 1.4 46.5nm
14.	eP	A	07 07 31.5	Guatemala 14.71 N 91.31 W
	epP	A	08 02	H = 06 55 01.8 h = 138 km MB = 5.3
				D = 86.75 Az = 38.4 (NEIS)
				h = 122 km
				PV A 1.6s 41.2nm M = 5.1

Day	Phase		h m s	Remarks
14.	ePg	A	07 16 26	Northern Italy 45.6 N 11.8 E
	eSg	A	17 28	H = 07 14 46 (BCIS)
				D = 5.07
14.	eP	A	10 56 08	Ryukyu Islands 25.99 N 128.49 E
	eX	A	56 17	H = 10 43 31.2 h = normal MB = 5.1
	eS	C	11 06 52	D = 85.58 Az = 325.3 (NEIS)
	eSSS	C	17.0	PV A 2.0s 42.8nm M = 5.3
	LmH	B	39.6	XV A 2.0 68.4nm
	LmV	B	39.7	LmH B 14 1.5 μm 5.5
				LmV B 14.5 1.6 μm 5.6
14.	e(P)	A	11 34 07	Crete 34.42 N 25.63 E
				H = 11 29 31.3 h = 2 km MB = 4.2
				D = 19.17 Az = 332.0 (NEIS)
14.	eX	A	11 51 07	Ryukyu Islands 25.95 N 128.46 E
	LmV	B	12 34.4	H = 11 38 21.3 h = normal MB = 5.1
	LmH	B	34.5	D = 85.60 Az = 325.3 (NEIS)
				XV A 2.0s 42.8nm
				LmH B 14.5 1.1 μm M = 4.4
				LmV B 14 1.4 μm 4.5
14.	eP	A	12 06 37.5	Dodecanese Islands 36.38 N 27.06 E
				H = 12 02 25.3 h = 45 km MB = 4.0
				D = 18.08 Az = 326.9 (NEIS)
14.	eP	A	15 53 50.5	Eastern Sibiria 64.20 N 174.00 W
				H = 15 43 10.56 h = normal MB = 4.5
				D = 65.41 Az = 356.1 (NEIS)
				PV A 1.2s 8.1nm M = 4.7
14.	e(P)	A	17 27 26	Ryukyu Islands 27 N 128 E
	LmH	B	18 03.3	H = 17 14 46 h = 89 km MB = 4.5
	LmV	B	10.8	D = 84.49 Az = 325 (ISC)
				(P)V A 2.0s 29.9nm M = 5.2
				LmH B 20 0.5 μm 4.9
				LmV B 14 0.6 μm 5.1

Day	Phase		h m s	Remarks
14.	ePKP	A	18 44 34.5	<u>Loyalty Islands</u> 20.90 S 168.51 E H = 18 24 58.3 h = normal MB = 5.0 D = 145.19 Az = 334.0 (NEIS) PKPV A 1.8s 30.4nm
14.	ePKP e	A A	19 48 51 49 11.5	<u>Loyalty Islands</u> 20.98 S 168.71 E H = 19 29 13.2 h = 25 km MB = 4.6 D = 145.33 Az = 334.2 (NEIS) PKPV A traces
14.	ePKP	A	22 08 02.5	<u>Loyalty Islands</u> 20.90 S 168.67 E H = 21 48 21.9 h = 4 km MB = 5.4 D = 145.25 Az = 334.2 (NEIS) PKPV A 1.6s 19.2nm
15.	eP LmH LmV	A C C	03 56 36 04 42.3 42.3	<u>Luzon, Philippine Islands</u> 18.91 N 120.83 E H = 03 43 52.4 h = 45 km MB = 5.3 MS=4.8 D = 87.13 Az = 323.0 (NEIS) PV A 1.7s 24.2nm M = 5.2 LmH C 17 1.1/um 5.3
15.	eSn eSg	A A	21 30 26 31 06	<u>Central Italy</u> 43.8 N 13.1 E H = 21 27 22 (BCIS) D = 6.95
15.	ePn ePg eSn eSg LmH LmV	A A A A B B	21 50 43 51 17 51 50 52 37 53.9 53.9	<u>Northern Italy</u> 44.53 N 9.85 E H = 21 49 13.6 h = normal MB = 4.5 D = 6.23 Az = 10.4 (NEIS) LmH B 10s 0.9/um M = 3.6 LmV B 10 1.2/um
16.	ePKP e e	A A A	11 11 09 11 29 11 38.5	<u>Loyalty Islands</u> 20.8 S 168.8 E H = 10 51 34 h = 33 km D = 145.19 Az = 334 (ISC) PKPV A 1.5s 20.1nm

Day	Phase		h m s	Remarks
16.	eP	A	11 35 45	Mindoro, Philippine Islands
	epP	A	36 13	13.83 N 120.65 E
	esP	A	36 23.5	H = 11 22 52.9 h = 123 km MB = 5.4
	LmH	C	12 13.0	D = 91.06 Az = 323.0 (NEIS)
				h = 109
				PV A 1.7s 51.5nm M = 5.4
				LmH C 15 0.3 μm
17.	eP	A	00 39 32	North Atlantic Ridge 35.18 N 35.34 W
	ePP	ABC	40 53	H = 00 32 21.4 h = normal MB=5.1 MS=5.0
	eS	BC	45 20	D = 37.04 Az = 50.6 (NEIS)
	LmH	B	52.3	PV A 2.5s 92.2nm M = 5.2
	LmV	B	54.2	SH B 15 1.1 μm 5.4
				LmH B 14.5 1.5 μm 4.9
				LmV B 14 1.4 μm 5.0
17.	eP	A	00 51 35.5	Andreanof Islands, Aleutian Is. 51.74 N 173.49 W
	epP	A	51 49	H = 00 39 40.7 h = 46 km MB = 4.9
				D = 77.90 Az = 356.7 (NEIS)
				PV A 1.2s 16.3nm M = 4.9
17.	e(Pn)	A	01 33 46	Rumania 45.97 N 21.14 E
	eSn	A	34 53.5	H = 01 31 33.9 h = normal MB = 5.6 (NEIS)
	eiSb	A	35 22	D = 7.88
17.	eP	A	05 57 19	Kurile Islands Region 49.77 N 157.10 E
				H = 05 45 35.2 h = normal MB = 4.6
				D = 75.71 Az = 338.1 (NEIS)
17.	eP	A	06 27 06	Kurile Islands 49.87 N 156.90 E
				H = 06 15 22.7 h = normal MB = 4.6
				D = 75.57 Az = 338.0 (NEIS)
17.	eP	A	06 29 22	Kurile Islands Region 49.74 N 157.03 E
				H = 06 17 38.1 h = normal MB = 4.5 (NEIS)
				D = 75.71
				PV A 1.0s 11.8nm M = 4.8

Day	Phase		h m s	Remarks
17.	eP	A	15 35 30	Molucca Passage 1.52 N 126.24 E H = 15 21 24.3 h = normal MB=5.5 MS=4.3 D = 104.18 Az = 323.3 (NEIS)
17.	eP	A	15 35 30	Molucca Passage 1.49 N 126.32 E H = 15 21 30.1 h = 86 km MB = 5.3 (ISC) D = 104.3
17.	-eP eS eSS LmH LmV	AB B B B B	18 35 10.5 41 20 44 08 56.4 59.3	Red Sea 17.26 N 40.37 E H = 18 27 33.7 h = normal MB=5.0 MS=5.1 D = 40.47 Az = 331.9 (NEIS) PV A 2.0s 120.0nm M = 5.3 SH B 12.5 1.0/um 5.6 LmH B 16 0.6/um 4.6 LmV B 12 0.5/um 4.7
18.	eSg	A	02 29 08.5	France 44.7 N 2.5 E H = 02 24 35 (BCIS) D = 8.59
18.	ePKP	A	02 59 29	Tonga Islands 16.05 S 174.25 W H = 02 40 09.0 h = 160 km MB = 4.1 D = 145.17 Az = 353.5 (NEIS) PKPV A 1.0s 13.8nm
18.	ePKP	A	08 24 24	Fiji Islands Region 17.72 S 178.30 W H = 08 05 42.8 h = 541 km MB = 5.0 D = 146.20 Az = 348.6 (NEIS) PKPV A 1.5s 15.1nm
18.	eP	A	10 42 36	Off East Coast of Kamchatka 53.90 N 163.56 E H = 10 31 06.2 h = 46 km MB=5.0 MS=4.2 D = 73.20 Az = 341.8 (NEIS) PV A 1.0s 31.5nm M = 5.2

Day	Phase		h m s	Remarks
18.	ePKP	A	14 35 22	Loyalty Islands 20.61 S 168.37 E
				H = 14 15 47.8 h = normal MB = 4.7
				D = 144.86 Az = 334.1 (NEIS)
				PKPV A 1.8s 33.8nm
18.	ePKP	A	16 23 07	Loyalty Islands 20.91 S 168.73 E
				H = 16 03 30.3 h = normal MB = 5.4
				D = 145.29 Az = 334.2 (NEIS)
				PKPV A 1.6s 22.0nm
18.	ePKIKP	A	20 52 06	West Chile Rise 38.26 S 93.77 W
	LmH	C	21 45.0	H = 20 33 01.4 h = normal MB=5.1 MS=4.7
				D = 127.48 Az = 50.7 (NEIS)
				PKIKPV A 1.6s 22.0nm
				LmH C 20 0.3/um M = 5.0
18.	ePKIKP	A	20 58 40	West Chile Rise 38.30 S 93.68 W
				H = 20 39 36.7 h = normal MB=5.2 MS=4.3
				D = 127.45 Az = 50.7 (NEIS)
				PKIKPV A 1.8s 20.3nm
18.	ePKIKP	A	21 25 01.5	West Chile Rise 38.17 S 93.67 W
	LmH	C	22 18.5	H = 21 05 58.5 h = normal MB=5.5 MS=4.7
				D = 127.37 Az = 50.6 (NEIS)
				PKIKPV A 1.7s 42.5nm
19.	ePKIKP	AB	07 23 49.5	South of Fiji Islands 24.01 S 178.51 E
	eiPKHKP	AB	23 57	H = 07 05 09.1 h = 596 km MB = 5.5
	eiPKP2	AB	24 09	D = 151.53 Az = 342.4 (NEIS)
	ep(PKHKP)	AB	26 14	PKIKPV A 1.7s 60.6nm
				PKHKPV A 1.5 151.0nm
				PKP2V A 1.5 186.0nm
19.	ePn	A	08 00 16	Czechoslovakia 50.76 N 14.42 E
	ePg	A	00 22	Explosion 18.3 t
	eiSg	A	00 42	H = 07 59.7 (PRU)
				D = 1.79 Az = 267 (ISC)

Day	Phase		h m s	Remarks
19.	ePKIKP	A	08 21 55	South of Fiji Islands 24.11 S 178.67 E
	ePKHKP	A	22 02	H = 08 03 13.3 h = 601 km MB = 5.1
	ePKP2	A	22 13.5	D = 151.67 Az = 342.5 (NEIS)
				PKIKPV A traces
				PKHKPV A 1.3s 24.0nm
				PKP2V A 1.2 32.5nm
19.	e	A	22 50 23	North Atlantic Ridge 21.06 N 45.76 W
				H = 22 41 02.3 h = 33 km MB = 4.1
				D = 53.41 Az = 42 (ISC)
				traces
20.	ePKHKP	A	01 46 49	Tonga Islands 19.46 S 173.14 W
	ePKP2	A	46 54	H = 01 27 02.7 h = normal MB = 4.7
				D = 148.66 Az = 354.2 (NEIS)
				traces
20.	ePKHKP	A	02 20 45	Loyalty Islands Region 22.07 S 171.85 E
				H = 02 01 00.7 h = normal MB=5.1 MS=4.8
				D = 148.28 Az = 335.8 (NEIS)
20.	ePKP2	A	03 29 58	Loyalty Islands Region 23.04 S 171.78 E
	LmV	C	04 56.0	H = 03 10 07.9 h = normal MB=5.2 MS=5.4
	LmH	C	56.4	D = 148.40 Az = 335.6 (NEIS)
				PKP2V A 2.0s 68.4nm
				LmH C 18 0.56 μm M = 5.3
				LmV C 18 0.30 μm 5.1
20.	eP	A	04 40 06	Philippine Islands Region
	ePP	A	43 50.5	13.75 N 125.07 E
				H = 04 26 51.5 h = normal MB = 5.1
				D = 93.68 Az = 324.2 (NEIS)
20.	e	A	05 24 41	
20.	eP	A	08 01 11	Nicobar Islands Region 6.97 N 94.73 E
				H = 07 49 00.5 h = normal MB = 4.6
				D = 80.28 Az = 320.1 (NEIS)

Day	Phase		h m s	Remarks
20.	ePKP	AB	08 46 59	Loyalty Islands Region 22.84 S 171.86 E
	LmV	B	09 56.0	H = 08 27 15.4 h = normal MB=5.1 MS=5.5
	LmH	B	56.8	D = 148.26 Az = 335.9 (NEIS)
				LmH B 18s 0.7 μm M = 5.4
				LmV B 19 0.7 μm 5.4
20.	ePKP	A	09 58 06	Loyalty Islands Region 22.98 S 171.96 E
				H = 09 38 18.8 h = normal MB = 4.9
				D = 148.42 Az = 335.9 (NEIS)
				traces
20.	eP	A	16 14 51.5	Hokkaido, Japan Region 42.36 N 142.99 E
				H = 16 02 58.3 h = 66 km MB = 5.1
				D = 78.13 Az = 330.8 (NEIS)
				PV A 1.0s 19.7nm M = 5.0
21.	ePKHKP	A	01 01 52	Fiji Islands Region 20.06 S 178.36 W
				H = 00 43 16.58 h = 667 km MB = 4.6
				D = 148.47 Az = 347.8 (NEIS)
				PKHKPV A 1.0s 15.8nm
21.	PKIKP	AB	01 13 13	Loyalty Islands Region 22.82 S 171.69 E
	ePKHKP	A	13 21	H = 00 53 30.0 h = normal MB=5.3 MS=5.6
	LmH	B	02 22.3	D = 148.17 Az = 335.7 (NEIS)
	LmV	B	22.5	PKIKPV A 1.0s 15.8nm
				PKHKPV A 2.3 115.8nm
				LmH B 21 1.3 μm M = 5.6
				LmV B 20 1.1 μm 5.6
21.	eP	A	01 36 42	Kurile Islands 43.50 N 146.70 E
				H = 01 24 45.2 h = 62 km MB = 4.7
				D = 78.41 Az = 332.7 (NEIS)
21.	eP	A	02 19 47	Sea of Okhotsk 46.25 N 145.36 E
	e	A	19 54.5	H = 02 08 02.9 h = 28 km MB=5.3 MS=5.5
				D = 75.54 Az = 331.6 (NEIS)
				PV A 1.4s 16.3nm M = 4.9

Day	Phase		h m s	Remarks
21.	e(P)	A	04 03 16	Southern Italy 39.0 N 17.1 E
	e(S)	A	05 22	H = 04 00 08 (BCIS)
				D = 12.32
21.	ePKHKP	A	05 40 33	Loyalty Islands Region 22.97 S 171.73 E
				H = 05 20 47.7 h = 47 km MB = 4.7
				D = 148.32 Az = 335.6 (NEIS)
21.	ePKP	A	11 53 59	Loyalty Islands Region 22.56 S 171.78 E
				H = 11 34 18.1 h = 55 km MB = 4.7
				D = 147.97 Az = 336.0 (NEIS)
				traces
21.	eP	A	15 19 22.5	Taiwan Region 25.25 N 122.14 E
				H = 15 07 20.1 h = 210 km MB = 4.9
				D = 82.82 Az = 323.1 (NEIS)
22.	eP	AB	00 41 07	Eastern China 31.65 N 119.17 E
	LmH	B	01 11.3	H = 00 29 19.8 h = normal MB=5.2 MS=5.5
	LmV	B	19.2	D = 76.18 Az = 321.3 (NEIS)
				PV A 2.0s 38.5nm M = 5.1
				LmH B 20 2.3 /um 5.5
				LmV B 14 1.2 /um 5.4
22.	ePKP	A	01 26 21	New Hebrides Islands 20.84 S 169.42 E
	e	A	26 31	H = 01 06 44.3 h = 22 km MB = 5.2
				D = 145.50 Az = 334.9 (NEIS)
				PKPV A 1.3s 13.1nm
22.	eP	A	01 50 27	Andaman Islands Region 14.17 N 93.98 E
				H = 01 38 51.6 h = normal MB = 4.9
				D = 74.35 Az = 319.1 (NEIS)
				PV A 1.2s 16.3nm M = 4.9
22.	ePKHKP	AB	02 25 07	Loyalty Islands Region 22.89 S 171.79 E
	e	A	25 32	H = 02 05 21.7 h = 38 km MB = 5.2

Day	Phase		h m s	Remarks
cont. 22.	LmH	C	03 34.0	D = 148.27 Az = 335.8 (NEIS)
	LmV	C	34.0	PKHKPV A 2.2s 43.5nm
				LmH C 20 0.24 µm M = 4.9
				LmV C 19 0.39 µm 5.2
22.	eP	A	03 24 39.5	**Eastern Mediterranean Sea**
				34.88 N 27.68 E
				H = 03 20 13.5 h = 65 km MB = 4.4
				D = 19.62 Az = 328.3 (NEIS)
				PV A 1.3s 24.0nm M = 4.3
22.	eP	A	04 34 22	**Fox Islands, Aleutian Is.**
				52.41 N 169.54 W
				H = 04 22 29.6 h = 43 km MB = 4.2
				D = 77.32 Az = 359.2 (NEIS)
22.	eP	A	04 42 53	**Fox Islands, Aleutian Is.**
				52.56 N 169.55 W
				H = 04 31 00.5 h = 32 km MB = 4.5
				D = 77.17 Az = 359.2 (NEIS)
22.	eP	A	07 29 25.5	**Near East Coast of Kamchatka**
	epP	A	29 40	53.90 N 160.48 E
				H = 07 18 00.7 h = 46 km MB = 4.6
				D = 72.59 Az = 339.8 (NEIS)
				h = 54 km
22.	eSn	A	12 28 13	**Austria** 47.29 N 16.71 E
				H = 12 26 12.0 h = 33 km (ISC)
				D = 4.7
22.	LmH	C	14 31.0	**Kurile Islands** 50.26 N 156.84 E
	LmV	C	31.0	H = 13 41 51.9 h = 59 km MB = 4.7 (ISC)
				D = 75.5
22.	eP	A	15 36(30)	**North Atlantic Ridge** 15.11 N 45.20 W
	LmH	C	16 03.0	H = 15 26 41.4 h = normal MB=4.6 MS=4.0

Day	Phase		h m s	Remarks
cont. 22.	LmV	C	16 05.0	D = 57.57 Az = 39.1 (NEIS)
				PV traces
				LmH C 15s 0.3/um M = 4.5
				LmV C 16 0.4/um 4.7
22.	eP	A	18 02 02	Greece 38.6 N 22.5 E
	e	A	02 50	H = 17 58 39 (BCIS)
	LmV	C	08.0	D = 14.33
	LmH	C	08.5	PV A traces
				LmH C 20s 0.20/um M = 2.3
				LmV C 12 0.47/um
22.	eP	A	21 00 42	Off East Coast of Honshu, Japan
				35.54 N 142.31 E
				H = 20 48 17.2 h = 58 km MB = 4.9
				D = 83.82 Az = 331.0 (NEIS)
23.	eP	A	17 48 31.5	Central Mid-Atlantic Ridge
	LmH	C	18 05.3	7.39 N 35.34 W
	LmV	C	09.3	H = 17 38 39.6 h = normal MB=4.7 MS=4.3
				D = 58.02 Az = 33.3 (NEIS)
				PV A 1.4s 11.6nm M = 4.8
				LmH C 18 0.35/um 4.5
24.	eiPKP	A	01 27 46	Tonga Islands 16.22 S 175.14 W
				H = 01 08 42.7 h = 306 km MB = 4.8
				D = 145.23 Az = 352.4 (NEIS)
				PKPV A 1.3s 43.7nm
24.	eP	A	03 05 04	Southern Iran 28.15 N 55.38 E
	e	A	05 18	H = 02 57 33.6 h = 62 km MB = 4.4
				D = 39.83 Az = 316.6 (NEIS)
				PV A traces
24.	ePg	A	06 59 26.5	Czechoslovakia 49.9 N 19.0 E
	eSg	A	07 00 29	H = 06 58 03 h = 0 km
				D = 4.78 Az = 281 (ISC)
				PgV A 0.8s 23.1nm

Day	Phase		h m s	Remarks
24.	e(Sg)	A	09 02 32.5	Czechoslovakia 50.63 N 15.67 E H = 09 00.7 (KHC) D = 2.58 Az = 272 (ISC)
24.	eP	A	24 08 30.5	Kurile Islands 47.92 N 154.17 E H = 23 56 44.1 h = 57 km MB = 4.9 D = 76.66 Az = 337 (ISC)
25.	eP	A	00 12 56	Uganda 1.00 N 30.09 E
	eSS	C	24 20	H = 00 03 49.1 h = normal MB = 5.0
	LmH	B	37.2	D = 51.90 Az = 345.1 (NEIS)
	LmV	B	39.9	LmH B 17s 0.4/um M = 4.5
				LmV B 16 0.4/um 4.6
25.	eP	A	16 01 07	Ryukyu Islands Region 29.82 N 131.85 E H = 15 48 37.2 h = normal MB = 5.0 D = 84.09 Az = 326.4 (NEIS) PV A 1.2s 12.2nm M = 4.9
26.	eP	A	02 30 00	Kurile Islands Region 44.57 N 151.75 E H = 02 17 53 h = 14 km MB = 4.3 D = 79.04 Az = 335 (ISC)
26.	ePg	A	07 22 37	Switzerland 47.67 N 7.10 E
	eSn	A	22 58	H = 07 21 14.2 h = normal
	eSg	A	23 27	D = 4.20 Az = 43.2 (NEIS)
26.	ePg	A	07 26 36	Switzerland 47.12 N 7.90 E
	eSg	A	27 27	H = 07 25 06.7 h = normal
				D = 4.30 Az = 33.5 (NEIS)
26.	eP	A	18 15 56	Red Sea 17.14 N 40.38 E H = 18 08 16.9 h = normal MB = 4.8 D = 40.58 Az = 331.9 (NEIS)
26.	eP	A	23 08 09	Hindu Kush Region 35.94 N 70.18 E H = 23 00 08.1 h = 80 km MB = 5.0 D = 43.94 Az = 308.5 (NEIS)

Day	Phase		h m s	Remarks
27.	eP	A	06 15 12	Southern Peru 14.96 S 72.21 W
	ePP	A	19 07.5	H = 06 01 47.3 h = 113 km MB = 5.8
				D = 97.57 Az = 39.7 (NEIS)
				traces
27.	ePKP	A	06 37 32	Fiji Islands Region 17.92 S 178.39 W
				H = 06 18 55.19 h = 579 km MB = 4.5
				D = 146.38 Az = 348.5 (NEIS)
				PKPV A 1.3s 17.5nm
27.	-ePKIKP	AB	07 44 43	South of Tonga Islands
	iPKHKP	A	44 53	26.25 S 175.91 W
	ePKP2	A	08 00 06	H = 07 24 54.0 h = 45 km MB=6.1 MS=5.9
	ePP	B	48 43	D = 154.96 Az = 348.6 (NEIS)
	ePPP	B	52 11	PKIKPV A 3.5s 958.3nm
	e	BC	55 20	PKHKPV B 6 2.5/um
	e	C	59 04	PPV B 8 1.6/um
	LmH	B	58.6	LmH B 18 2.5/um M = 6.0
	LmV	B	09 07.7	LmV B 17 3.3/um 6.2
27.	eP	A	10 12 20.5	Near East Coast of Kamchatka
	epP	A	12 30.5	55.91 N 162.86 E
				H = 10 01 02.2 h = 26 km MB=5.0 MS=4.9
				D = 71.16 Az = 341.1 (NEIS)
				h = 33 km
				PV A 1.2s 24.4nm M = 5.2
27.	ePKHKP	A	10 21 47	South of Fiji Islands 25.14 S 179.79 E
				H = 10 02 48.1 h = 519 km MB = 4.8
				D = 152.95 Az = 343.3 (NEIS)
28.	eP	A	01 00 10	Turkey 36.27 N 31.85 E
				H = 00 55 32.8 h = normal MB = 4.2
				D = 20.46 Az = 320.9 (NEIS)
				PV A 1.2s 16.3nm M = 4.3

Day	Phase		h m s	Remarks
28.	LmH	C	05 00.8	<u>Northern Easter Island Cordillera</u>
	LmV	C	02.0	3.88 S 104.06 W
				H = 03 57 49.8 h = 33 km MB = 5.0 (ISC)
				D = 109.3
				LmH C 16s 0.45/um M = 5.1
				LmV C 19 0.55/um 5.1
28.	ePKHKP	A	12 56 12	<u>Fiji Islands Region</u> 20.78 S 177.17 W
	epPP	A	13 01 12.5	H = 12 37 00.7 h = 319 km MB = 4.7
	esPP	A	01 49	D = 149.40 Az = 349.0 (NEIS)
	eSKS	A	02 38	PKHKPV A 1.2s 10.2nm
28.	ePKHKP	A	17 19 51	<u>Loyalty Islands Region</u> 22.97 S 171.77 E
	LmV	C	18 40.0	H = 17 00 07.5 h = 53 km MB = 4.9
				D = 148.34 Az = 336 (ISC)
				LmV C 19s 0.35/um
28.	LmH	C	20 09.1	LmH C 19s 0.35/um
	LmV	C	09.2	LmV C 19 0.45/um
29.	+eP	AB	20 10 02.5	<u>United Arab Republic</u> 30.53 N 31.72 E
	LmH	C	23.6	H = 20 04 39.7 h = normal MB = 4.9
				D = 25.09 Az = 328.9 (NEIS)
				PV A 0.9s 62.3nm M = 5.2
				LmH C 21 0.35/um 3.8
29.	ePKP	A	21 13 39	<u>Fiji Region</u> 14.2 S 179.6 W
				H = 20 54 14 h = 0 km
				D = 142.50 Az = 348 (ISC)
29.	eP	A	22 33 19.5	<u>Colombia</u> 4.75 N 76.15 W
				H = 22 20 52.2 h = 87 km MB = 5.1
				D = 84.94 Az = 39.7 (NEIS)
				traces
30.	ePKHKP	A	19 49 54	<u>Fiji Islands Region</u> 20.29 S 177.92 W
				H = 19 31 07.4 h = 550 km MB = 4.3
				D = 148.78 Az = 348.3 (NEIS)
				PKHKPV A 1.6s 38.5nm

Day	Phase		h m s	Remarks
30.	e	A	19 59 51	Andreanof Islands, Aleutian Is.
	e	A	59 58.5	51.10 N 172.73 W
				H = 19 47 39.8 h = 12 km MB = 4.8
				D = 78.57 Az = 357.2 (NEIS)
30.	ePKP	A	20 15 09	New Hebrides Islands 19.66 S 169.21 E
	epPKP	A	15 42	H = 19 55 52.1 h = 158 km MB = 5.0
				D = 144.35 Az = 335.4 (NEIS)
				h = 121 km
				PKPV A 1.2s 20.3nm

Day	Phase		h m s	Remarks
1.	ePKHKP	A	05 17 56	South of Tonga Islands 26.34 S 175.60 W
	ePKP2	A	18 11	H = 04 57 57.7 h = 44 km MB=5.0 MS=4.5
				D = 155.11 Az = 349.0 (NEIS)
1.	LmH	C	13 44.3	South Sandwich Islands Region
				58.84 S 25.37 W
				H = 12 43 49 h = 49 km MB = 5.1 (ISC)
				D = 113.0
				LmH C 24s 0.4/um M = 4.9
1.	eP	A	15 35 21.5	Mariana Islands 18.28 N 145.18 E
				H = 15 22 24.7 h = 455 km MB = 5.5
				D = 100.17 Az = 332.0 (NEIS)
				PV A 1.8s 47.4nm M = 5.6
1.	ePKIKP	A	18 54 38	South of Fiji Islands 23.85 S 179.85 E
	ePKHKP	A	54 46	H = 18 35 51.2 h = 522 km MB = 4.8
	ePKP2	A	54 58.5	D = 151.73 Az = 344.1 (NEIS)
	epPKP	A	56 55	PKIKPV A traces
				PKHKPV A 1.3s 17.5nm
1.	eP	A	22 07 46.5	Near East Coast of Kamchatka
				53.62 N 160.30 E
				H = 21 56 19.4 h = normal MB = 4.7
				D = 72.83 Az = 339.7 (NEIS)
				PV A 1.1s 10.1nm M = 4.7
2.	eP	A	04 17 24	Near East Coast of Honshu, Japan
	LmH	C	56.0	35.02 N 141.26 E
				H = 04 04 55.3 h = 42 km MB = 4.8
				D = 83.85 Az = 330.5 (NEIS)
				LmH C 18s 0.7/um M = 5.1
2.	eP	A	05 46 46	Near East Coast of Honshu, Japan
				35.14 N 141.38 E
				H = 05 34 19.5 h = normal MB = 4.6
				D = 83.80 Az = 330.5 (NEIS)

Day	Phase		h m s	Remarks
2.	eP	A	05 48 00.5	Near East Coast of Honshu, Japan
	LmH	C	06 29.6	35.25 N 141.30 E
				H = 05 35 31.8 h = 14 km MB = 4.9
				D = 83.67 Az = 330.5 (NEIS)
				LmH C 16s 0.8nm M = 5.2
2.	ePKHKP	A	17 18 35.5	South of Fiji Islands 24.24 S 176.20 W
	ePKP2	A	18 50	H = 16 58 55.9 h = 171 km MB = 4.9
				D = 152.95 Az = 349.0 (NEIS)
				traces
2.	eP	A	21 45 08.5	Near East Coast of Honshu, Japan
	LmH	B	22 23.6	35.09 N 141.30 E
	LmV	B	27.5	H = 21 32 41.3 h = normal MB = 5.0
				D = 83.80 Az = 330.5
				PV A 1.6s 30.2nm M = 5.2
				LmH B 16 0.7 µm 5.2
				LmV B 15 0.3 µm 4.9
2.	eP	A	22 47 49.5	Off East Coast of Kamchatka
				51.62 N 159.49 E
				H = 22 36 15.3 h = 59 km MB = 4.6
				D = 74.53 Az = 339.4 (NEIS)
3.	e	A	12 41 21	
3.	eP	A	14 09 48	Hindu Kush Region 36.16 N 69.85 E
				H = 14 01 56.0 h = 153 km MB = 4.6
				D = 43.60 Az = 308.3 (NEIS)
				traces
3.	LmH	C	23 28.7	Near East Coast of Honshu
				35.04 N 141.34 E
				H = 22 35 04.5 h = 41 km MB = 4.4
				or
				H = 22 40 28.4 35.01 N 141.29 E
				MB = 4.4 (ISC)
				D = 83.9
				LmH C 20s 0.4 µm M = 4.8

Day	Phase		h m s	Remarks
4.	ePn	A	00 21 10	Northern Italy 45.9 N 11.6 E
	e	A	21 16	H = 00 19 46 h = 0 km
	e	A	21 49	D = 4.75 Az = 359 (ISC)
	eSn	A	21 55.5	
	e(Sb)	A	22 16.5	
4.	ePKIKP	A	01 37 17	New Britain Region 4.86 S 152.50 E
				H = 01 18 24.8 h = 69 km MB = 5.3
				D = 123.86 Az = 331.1 (NEIS)
				traces
4.	eP	A	07 31 47.5	New Ireland Region 4.70 S 153.21 E
				H = 07 12 54.7 h = 70 km MB = 5.0
				D = 124.05 Az = 331.5 (NEIS)
4.	eP	A	08 13 30	Kodiak Island Region 56.30 N 153.27 W
	e	A	13 38	H = 08 02 00.2 h = 10 km MB = 4.6
				D = 72.69 Az = 10.0 (NEIS)
4.	eP	A	08 58 36	Kodiak Island Region 56.30 N 153.29 W
	e	A	58 44	H = 08 47 04.6 h = 2 km MB = 4.4
				D = 72.69 Az = 10.0 (NEIS)
4.	ePKIKP	A	09 28 50.5	South of Fiji Islands 24.76 S 178.94 E
	ePKHKP	A	28 57.5	H = 09 10 01.9 h = 545 km MB = 5.2
	ePKP2	A	29 10.5	D = 152.36 Az = 342.5 (NEIS)
				PKIKPV A 1.3s 24.0nm
				PKHKPV A 1.3 69.9nm
				PKP2V A 1.4 105.0nm
4.	ePKIKP	AB	13 05 45	New Hebrides Islands Region
	eX	A	05 52.5	13.88 S 172.61 E
	epPKIKP	BC	08 05	H = 12 47 28.3 h = 602 km MB = 5.5
	ePP	BC	08 56	D = 140.21 Az = 341.1 (NEIS)
	eSS	C	26 40	PKIKPV A 1.4s 32.6nm
	eSSS	C	31 50	PKIKPV B 6 1.1 μm
	LmH	B	14 02.0	LmH B 16 1.0 μm
	LmV	C	02.5	LmV C 19 0.5 μm

Day	Phase		h m s	Remarks
4.	eP	A	15 18 02	Algeria 34.89 N 5.05 E
	e	A	18 06	H = 15 14 11.9 h = normal MB = 4.3
	e	A	18 10	D = 16.54 Az = 14.8 (NEIS)
	LmH	C	24.0	LmH C 16s 0.8 µm M = 4.0
4.	eP	A	18 06 51	South of Panama 7.71 N 82.63 W
				H = 17 54 07.0 h = normal MB = 5.2
				D = 86.79 Az = 39.5 (NEIS)
				PV A 1.6s 16.5nm M = 5.0
4.	eP	A	22 08 48	Off East Coast of Kamchatka
				51.58 N 159.67 E
				H = 21 57 10.0 h = normal MB = 4.6
				D = 74.61 Az = 339.6 (NEIS)
				PV A 1.0s 9.8nm M = 4.8
5.	eSg	A	04 49 57.5	Poland 50.36 N 18.90 E
				H = 04 47 16.9 M = 3.0 (WAR)
				D = 4.7
5.	eP	A	06 10 09	Taiwan Region 22.30 N 121.49 E
	ePP	A	13 15	H = 05 57 35.1 h = 26 km MB = 5.5
	LmV	B	53.1	D = 84.81 Az = 323.0 (NEIS)
	LmH	B	53.2	PV A 1.3s 56.8nm M = 5.6
				PPV A 1.3 26.2nm 5.6
				LmH B 16 1.7 µm 5.5
				LmV B 16 2.3 µm 5.7
5.	LmV	B	06 20.3	Algeria 35.04 N 4.53 E
	LmH	B	21.7	H = 06 09 22 h = 53 km MB = 4.3 (ISC)
				D = 16.5
				LmH B 10s 0.7 µm
				LmV B 14 0.3 µm
5.	ePKIKP	AC	08 37 14	New Hebrides Islands 17.45 S 167.91 E
	ePP	AC	40 26	H = 08 17 50.3 h = 33 km MB=5.1 MS=5.6
	LmV	B	09 47.7	D = 141.84 Az = 335.5 (NEIS)
	LmH	B	48.8	LmH B 18.5s 0.7 µm M = 5.4
				LmV B 18 0.8 µm 5.5

Day	Phase		h m s	Remarks
5.	eP	AC	14 31 26.5	<u>Near East Coast of Honshu, Japan</u>
	epP	A	31 38	37.74 N 141.71 E
	ePP	A	34 27	H = 14 19 12.0 h = 48 km MB = 5.6
	epPP	A	34 44	D = 81.67 Az = 330.5 (NEIS)
	LmH	B	15 06.2	h = 45 km
	LmV	B	12.9	PV A 1.3s 65.5nm M = 5.5
				LmH B 16 0.6 µm 5.0
				LmV B 15 0.7 µm 5.2
5.	eP	A	19 22 23	<u>Kurile Islands</u> 46.15 N 149.22 E
	epP	A	22 54.5	H = 19 10 40.0 h = 143 km MB = 5.1
	esP	A	23 15	D = 76.85 Az = 333.8 (NEIS)
	LmH	C	55	h = 139 km
				PV A 1.5s 22.6nm M = 4.7
6.	-iPn	A	07 51 28	<u>Austria</u> 46.34 N 13.40 E
	ePg	A	51 45	H = 07 50 22.1 h = 54 km
	eSg	AB	52 41.5	D = 4.47 Az = 345.3 (NEIS)
6.	eX	A	10 45 17	<u>Northern Sumatra</u> 3.16 N 96.27 E
	LmH	B	11 20.7	H = 10 32 33.8 h = normal MB=4.9 MS=5.2
	LmV	B	35.7	D = 84.18 Az = 320.4 (NEIS)
				XV A 1.8s 40.5nm
				LmH B 18 1.1 µm M = 5.3
				LmV B 16 0.9 µm 5.3
6.	ePKP	A	11 57 56	<u>Tonga Islands</u> 15.20 S 173.42 W
				H = 11 38 19.7 h = 13 km MB=5.7 MS=5.2
				D = 144.41 Az = 354.5 (NEIS)
				PKPV A 1.3s 35.0nm
6.	ePKP2	A	20 35 26	<u>Tonga</u> 19.88 S 174.0 W
				H = 20 15 31.6 h = 33 km
				D = 148.98 Az = 353 (ISC)
				PKP2V A 1.6s 22.0nm

Day	Phase		h m s	Remarks
6.	LmV	B	23 17.3	Northeastern China 40.17 N 119.64 E
	LmH	B	17.4	H = 22 31 53.3 h = normal MB = 4.4
				D = 69.85 Az = 319.8 (NEIS)
				LmH B 15.5s 0.6 /um M = 5.9
				LmV B 16 0.6 /um 5.0
7.	e(Sg)	A	00 03 28.5	Yugoslavia 46.6 N 15.2 E
				(VIE)
				D = 4.68
7.	ePKP	A	02 44 49	Fiji Islands Region 16.69 S 177.34 W
	eSS	C	03 06 50	H = 02 25 10.8 h = normal MB=5.5 MS=6.0
	LmV	B	51.2	D = 145.37 Az = 350.0 (NEIS)
	LmH	B	51.3	PKPV A 1.9s 121.0nm
				PKPV B 4 0.45 /um
				LmH B 19 3.8 /um M = 6.1
				LmV B 19 3.0 /um 6.1
7.	eP1	A	03 14 56	North Atlantic Ridge 12.73 N 44.48 W
	eP2	A	14 59.5	H = 03 04 57.5 h = normal MB=5.2 MS=5.8
				D = 58.98 Az = 38.1 (NEIS)
				P1V A 1.4s 27.9nm M = 5.2
				P2V A 1.3 43.6nm 5.4
7.	LmH	B	12 40.1	Panama-Costa Rica 8.42 N 82.72 W
	LmV	B	40.8	Border Region
				H = 11 54 36.4 h = 72 km MB = 4.8 (ISC)
				D = 86.3
7.	eP	A	21 15 35	Kurile Islands Region 47.56 N 156.26 E
				H = 21 03 41.5 h = normal MB = 4.7
				D = 77.55 Az = 337.8 (NEIS)
8.	eP	A	04 37 56	Central Alaska 63.67 N 150.73 W
				H = 04 27 13.1 h = 11 km MB = 4.6
				D = 65.19 Az = 12.3 (NEIS)
				PV A 0.8s 13.5nm M = 5.2

Day	Phase		h m s	Remarks
8.	eP	A	11 10 47.5	Sea of Okhotsk 47.92 N 145.79 E
				H = 10 59 56.7 h = 453 km MB = 4.1
				D = 74.21 Az = 331.7 (NEIS)
				PV A 1.4s 14.0nm M = 4.4
8.	ePKHKP	A	12 10 59	Fiji Islands Region 17.95 S 178.34 W
				H = 11 52 26.5 h = 634 km MB = 4.5
				D = 146.43 Az = 348.5 (NEIS)
8.	e	A	14 44 50	
8.	eP	A	22 36 42.5	Molucca Passage 2.43 N 126.03 E
				H = 22 22 48.8 h = 87 km MB = 5.2
				D = 103.33 Az = 323.4 (NEIS)
8.	eP	AB	23 45 54.5	Near S. Coast of Honshu, Japan
	Pm	AB	46 19	34.52 N 138.74 E
	ePP	AB	49 11	H = 23 33 25.2 h = 2 km
	eS	B	56 18	MB = 6.0 MS = 6.5 (NEIS)
	eSS	B	24 01 40	D = 83.24
	LmH	B	23.7	PV A 1.8s 57.5nm M = 5.5
	LmV	B	27.7	PmV A 1.8 297.3nm 6.2
				PmV B 10 1.9 μm 6.3
				PPV A 3.0 657.9nm 6.5
				PPV B 4 1.5 μm 6.4
				PPH B 3.7 1.0 μm 6.8
				SH B 14.3 6.3 μm 6.6
				LmH B 16.5 110.0 μm 7.3
				LmV B 13.5 51.2 μm 7.1
9.	e	A	13 41 54	Molucca Passage 0.78 N 125.87 E
	ePP	A	46 09	H = 13 27 37.0 h = 18 km MB=5.4 MS=5.3
	eS	C	53 30	D = 104.55 Az = 323.2 (NEIS)
	LmH	B	14 31.4	LmH B 21s 1.2 μm M = 5.4
	LmV	B	33.4	LmV B 20 1.2 μm 5.4

Day	Phase		h m s	Remarks
9.	ePKIKP	AB	16 27 20.5	<u>Loyalty Islands Region</u> 21.76 S 169.75 E
	ePKHKP	A	27 22.5	H = 16 07 43.0 h = 35 km MB = 5.5
				D = 146.45 Az = 334.6 (NEIS)
				PKIKPV A 1.5s 60.3nm
9.	eP	A	17 06 40	<u>Dodecanese Islands</u> 36.60 N 27.64 E
				H = 17 02 21.7 h = normal MB = 4.5 (NEIS)
				D = 18.15 Az = 134.53
9.	ePP	C	24 14 12	<u>Prince Edward Islands Region</u>
	ePPPP	C	18 05	45.98 S 35.27 E
	eSKS	C	20 55	H = 23 56 38.0 h = normal MB=5.7 MS=5.9
	eiPS	C	23 15	D = 98.40 Az = 345.0 (NEIS)
	eSS	C	28 25	PPV C 8s 1.2 µm M = 6.3
	eSSSS	C	36 00	LmH B 17 8.8 µm 6.3
	LmV	B	55.2	LmV B 18 11.0 µm 6.4
	LmH	B	55.3	
10.	ePKIKP	A	02 23 15.5	<u>Kermadec Islands Region</u> 30.72 S 179.55 W
	ePKHKP	A	23 28.5	H = 02 03 46.1 h = 227 km MB = 5.3
	iPKP2	A	23 53.5	D = 158.42 Az = 340.4 (NEIS)
				PKIKPV A 1.5s 25.1nm
				PKP2V A 1.2 95.5nm
10.	eP	A	05 38 12.5	<u>Ryukyu Islands</u> 27.49 N 129.62 E
				H = 05 25 37.3 h = normal MB = 5.0
				D = 84.92 Az = 325.6 (NEIS)
10.	eP	A	06 32 14	<u>Near East Coast of Honshu, Japan</u>
				40.20 N 141.44 E
				H = 06 20 16.3 h = 93 km MB = 4.7
				D = 79.42 Az = 330.2 (NEIS)
10.	eP diff	B	08 26 28	<u>Northern Easter I. Cordillera</u>
	ePP	AB	30(55)	4.37 S 102.11 W
	ePS	C	40 10	H = 08 12 05.0 h = normal MB=6.1 MS=6.0
	ePPS	C	41 15	D = 108.29 Az = 37.9 (NEIS)
	e	C	45 32	LmH B 20s 2.0 µm M = 5.7

Day	Phase		h m s	Remarks
cont. 10.	eSS	C	08 46 15	LmV B 22s 2.5 um M = 5.7
	eSSS	C	50 10	
	LmH	B	09 10.5	
	LmV	B	10.5	
10.	e	A	11 00 44	
	e	A	01 03.5	
10.	eP1	A	19 36 30	Szechwan Province, China
	eP2	AB	36 32	28.24 N 104.02 E
	eS	B	45 48	H = 19 25 15.0 h = 11 km MB=6.2 MS=6.8
	eSS	C	49 56	D = 70.15 Az = 317.4 (NEIS)
	eiSSS	C	53 25	P1V A 1.6s 27.5nm M = 5.2
	LmH	B	20 03.8	P2V A 2.0 200.9nm 6.0
	P'P'	A	04 39	P2V B 5.5 1.6 um 6.4
	LmV	B	12.5	SH B 13 4.6 um 6.5
				LmH B 22 121.0 um 7.1
				P'P'V A 2.8 182.4nm
				LmV B 16 36.4 um 6.8
11.	eP	AC	00 57 48	Molucca Passage 1.74 N 126.38 E
	ePP	C	01 02 08	H = 00 43 44.9 h = normal MB=6.0 MS=6.0
	eSKS	C	08 24	D = 104.08 Az = 323.4 (NEIS)
	eS	C	09 34	PV A 1.6s 49.5nm M = 6.1
	ePS	C	10 56	LmH B 18.5 5.9 um 6.2
	eSS	C	17.5	LmV B 18 6.2 um 6.2
	LmH	B	49.9	
	LmV	B	49.9	
11.	ePKP	A	02 36 55	Fiji Islands Region 17.51 S 176.96 W
				H = 02 18 00.0 h = 418 km MB = 4.8
				D = 146.24 Az = 350.2 (NEIS)
11.	-iP	AB	06 27 55.3	Mariana Islands Region 19.74 N 147.27 E
	-iPP	AB	32 01	H = 06 14 08.7 h = 6 km MB=6.4 MS=5.9
	eSKS	C	38 40	D = 99.79 Az = 333.2 (NEIS)
	ePS	C	40 52	PV A 1.5s 95.5nm M = 6.1
	eSS	C	46 10	PPV A 1.6 330.0nm 6.5

Day	Phase		h m s	Remarks
cont. 11.	LmH	B	07 09.2	PPH A 1.7s 263.2nm M = 6.6
	LmV	B	18.3	PPV B 8 2.2/um 6.6
				LmH B 18 3.8/um 5.9
				LmV B 16.5 2.9/um 5.9
11.	eP	A	09 22 46	<u>Iceland</u> 64.95 N 20.93 W
				H = 09 17 49.2 h = 16 km MB = 4.6
				D = 22.14 Az = 114.7 (NEIS)
				PV A 1.5s 35.2nm M = 4.6
11.	LmH	B	14 02.7	<u>South Western Atlantic Ocean</u>
	LmV	B	03.4	58.15 S 7.17 W
				H = 13 01 05.6 h = 28 km MB = 5.5 (ISC)
				D = 109.6
				LmH B 18s 1.4/um M = 5.6
				LmV B 18 1.8/um 5.7
11.	eP	A	19 22 14	<u>Kodiak Islands Region</u> 56.50 N 153.08 W
				H = 19 10 47.8 h = 27 km MB = 4.9
				D = 72.47 Az = 10.2 (NEIS)
				PV A 0.8s 15.4nm M = 5.1
11.	eP	A	21 07 19	<u>Molucca Passage</u> 1.88 N 126.42 E
	ePP	A	11 37.5	H = 20 53 16.0 h = normal MB=5.5 MS=5.4
	eSKS	C	17 52	D = 104.00 Az = 323.4 (NEIS)
	eS	C	19 00	PV A 1.8s 33.78 nm M = 5.9
	eSP	C	20 36	PPV A 1.6 16.5nm 5.6
	LmH	B	59.3	LmH B 20 1.27/um 5.5
	LmV	B	59.6	LmV B 16 1.06/um 5.5
12.	eP	A	00 24 56	<u>Dodecanese Islands</u> 36.66 N 27.03 E
	e	A	26 05	H = 00 20 56.9 h = 156 km MB = 4.5
	e	A	26 28	D = 17.83 Az = 326.4 (NEIS)
				PV A 1.9s 60.6nm M = 4.6
12.	eP	A	10 19 26	<u>Northern Chile</u> 19.56 S 69.05 W
	ePP	AB	23 24	H = 10 05 55.4 h = 112 km MB = 5.8
	eSKS	B	29 58	D = 99.16 Az = 39.5 (NEIS)

Day	Phase		h m s	Remarks
cont. 12.	LmV	B	10 32.4	PPV A 1.8s 60.8nm M = 5.9
	LmH	B	32.7	LmH B 18 0.5/um
				LmV B 18 0.5/um
12.	LmV	B	13 01.0	<u>Molucca Passage</u> 2.80 N 126.77 E
	LmH	B	01.5	H = 12 00 48.6 h = 33 km MB = 5.3 (ISC)
				D = 103.5
				LmH B 18s 0.5/um M = 5.1
				LmV B 20 0.4/um 5.0
12.	iPn	A	19 48 58	<u>Federal Republic of Germany</u>
	iPg	A	49 07	48.27 N 9.10 E
	iSn	A	49 30.5	H = 19 48 13.3 h = 12 km
	eiSg	A	49 45	D = 2.89 Az = 33.6 (NEIS)
13.	eP	A	02 23 57	<u>Leyte, Philippine Islands</u>
				10.08 N 124.12 E
				H = 02 11 29.7 h = 592 km MB = 5.3
				D = 96.08 Az = 323.7 (NEIS)
				PV A 1.5s 50.3nm M = 5.6
13.	ePKIKP	A	12 12(00)	<u>Solomon Islands</u> 7.30 S 155.50 E
				H = 11 52 55.9 h = 33 km MB=5.6 MS=5.4
				D = 127.42 Az = 331.8 (NEIS)
13.	+eP	AB	17 48 17.5	<u>Hindu Kush Region</u> 36.47 N 70.91 E
	epP	A	49 02	H = 17 40 28.4 h = 208 km MB = 5.5
	esP	B	49 22	D = 44.08 Az = 308.1 (NEIS)
	ePP	B	51 10	h = 206 km
				PV A 1.8s 274.0nm M = 5.4
				PPV B 10 0.9/um 5.4
				LmH B 9.8 0.8/um
13.	eP	A	19 07 57	<u>Southwest of Sumatra</u> 6.67 S 102.65 E
	e	A	08 05	H = 18 54 32.2 h = normal MB = 5.6
	ePP	C	12 05	D = 95.76 Az = 320.2 (NEIS)
	eSKS	C	18 30	PV A 1.8s 23.7nm M = 5.4
	eS	C	19 10	LmH B 16 1.8/um 5.6

Day	Phase		h m s	Remarks
cont. 13.	LmH	B	20 03.0	LmV B 14.5s 2.1 μm M = 5.8
	LmV	B	03.1	
13.	eP	A	20 16 30	Molucca Passage 2.71 N 126.54 E H = 20 02 29.9 h = normal MB = 5.4 D = 103.40 Az = 323.6 (NEIS) PV A traces
14.	ePKIKP	A	06 28 47	Solomon Islands 6.90 S 155.25 E H = 06 09 45.0 h = normal MB = 5.3 D = 126.94 Az = 331.8 (NEIS)
14.	e	A	08 35 36	Western Russia, poorly determined 57.1 N 21.3 E H = 08 32 22 h = 0 km (ISC) D = 8.63
15.	eP	A	00 21 18	Kashmir-Sinkiang Border Region 36.38 N 76.48 E H = 00 12 46.4 h = 62 km MB = 5.0 D = 47.66 Az = 308.8 (NEIS)
15.	eP	A	04 01 58	India-East Pakistan Border Region 25.55 N 91.80 E H = 03 51 21.4 h = 32 km MB = 4.5 D = 64.61 Az = 316.0 (NEIS)
15.	eP	A	05 44 50	North Atlantic Ridge 27.41 N 44.23 W H = 05 36 12.7 h = normal MB=4.7 MS=4.4 D = 47.87 Az = 45.3 (NEIS) PV A traces
15.	ePKP	A	09 51 17	Samoa Islands Region 15.71 S 172.08 W H = 09 31 41.9 h = normal MB = 4.8 D = 145.03 Az = 355.9 (NEIS)

Day	Phase		h m s	Remarks
15.	eP	A	10 42 36	North Atlantic Ridge 27.38 N 44.23 W
	LmH	C	58.5	H = 10 33 58.9 h = normal MB = 5.0
				D = 47.89 Az = 45.3 (NEIS)
				PV A 1.5s 22.6nm M = 5.0
				LmH C 20 0.3/um 4.3
15.	eP	A	11 16 45	North Atlantic Ridge 27.34 N 44.43 W
				H = 11 08 07.3 h = 33 km MB = 4.6
				D = 48.04 Az = 45 (ISC)
				PV A 1.8s 13.5nm M = 4.7
15.	eP	A	13 15 56	Fox Islands, Aleutian Is.
				52.41 N 168.82 W
				H = 13 04 04.1 h = 44 km MB=5.0 MS=4.5
				D = 77.32 Az = 359.7 (NEIS)
				PV A 1.1s 38.3nm M = 5.3
15.	eP	A	13 45 49	North Atlantic Ridge 27.40 N 44.32 W
	LmV	B	14 02.6	H = 13 37 11.1 h = normal MB=5.0 MS=5.0
	LmH	B	03.0	D = 47.93 Az = 45.3 (NEIS)
				LmH B 17.5s 0.5/um M = 4.6
				LmV B 17 0.6/um 4.7
15.	eP	A	13 52 49	North Atlantic Ridge 27.37 N 44.37 W
				H = 13 44 10.5 h = normal MB = 4.6
				D = 47.99 Az = 45.3 (NEIS)
15.	eP	A	14 07 53	North Atlantic Ridge 27.31 N 44.42 W
	LmV	B	24.8	H = 13 59 15.2 h = normal MB=4.9 MS=4.8
	LmH	B	25.4	D = 48.05 Az = 45.2 (NEIS)
				PV A traces
				LmH B 16s 0.3/um M = 4.4
				LmV B 20 0.7/um 4.7
15.	+iP	ABC	19 11 34	Kurile Islands 50.05 N 156.11 E
	ePP	C	14 24	H = 18 59 55.9 h = 56 km MB = 6.1
	ePPP	C	16 36	D = 75.20 Az = 337.5 (NEIS)
	eiS	C	21 08	PV A 2.4s 622.0nm M = 6.1

Day	Phase		h m s	Remarks
cont. 15.	ePS	C	19 22 30	PV B 12s 3.5 µm M = 6.2
	eSS	C	26 58	LmH B 17 38.6 µm
	eSSS	C	30 00	LmV B 18 34.1 µm
	LmH	B	49.0	
	LmV	B	49.2	
15.	eP	A	19 38 09	North Atlantic Ridge 27.40 N 44.43 W
				H = 19 29 32.3 h = normal MB = 5.2
				D = 48.00 Az = 45.3 (NEIS)
				PV A 1.0s 9.8nm M = 4.8
15.	e	A	20 34 31	Iceland Region 66.4 N 18.3 W
				H = 20 29 28 h = 0 km MB = 4.0
				D = 21.90 Az = 122 (ISC)
15.	eP	A	22 30 17	North Atlantic Ridge 27.35 N 44.28 W
				H = 22 21 41 h = 44 km MB = 4.7
				D = 47.93 Az = 45 (ISC)
16.	+iP	A	03 10 46.5	Eastern Kazakh SSR 49.74 N 78.15 E
	ePn	A	12 19	H = 03 02 57.3 h = 0 km MB = 5.3
	e	A	12 23	D = 41.30 Az = 297.8 (NEIS)
				Underground explosion MB = 6.0 (UPP)
				PV A 0.8s 42.3nm M = 5.2
16.	ePKP	A	C5 38 29.5	New Hebrides 18.50 S 169.03 E
				H = 05 19 23.9 h = 225 km MB = 4.4 (NEIS)
				D = 143.23
				PKPV A 1.5s 20.1nm
16.	eP	AC	15 11 42	Dodecanese Islands 36.0 N 27.2 E
	eS	C	15 18	H = 15 07 26 (BCIS)
	LmH	C	17.0	D = 18.45
	LmV	C	19.7	LmV C 10s 0.90 µm M = 4.5
16.	-iP	AC	20 12 11.3	Bonin Islands Region 27.16 N 140.07 E
	ePP	AC	15 51	H = 20 00 01.5 h = 471 km MB = 5.3
	LmV	B	56.4	D = 90.15 Az = 330.1 (NEIS)

Day	Phase		h m s	Remarks
cont. 16.	LmH	B	20 58.5	PV A 2.1s 230.0nm M = 5.7
				PPV A 2.0 162.0nm 5.8
				LmH B 15.5 0.55 /um
				LmV B 18 0.6 /um
16.	eP	A	23 20 38	Near Coast of Nicaragua
	eX	A	20 58	11.24 N 86.05 W
	ePP	A	24 08	H = 23 07 46.6 h = 36 km MB=5.5 MS=5.6
	LmH	B	58.2	D = 86.21 Az = 39.2 (NEIS)
	LmV	B	58.3	PV A 1.5s 15.1nm M = 5.0
				XV A 1.8 43.9nm
				LmH B 18 4.2 /um 5.9
				LmV B 18.5 4.9 /um 6.0
17.	ePKP	A	09 46 52.5	Tonga Islands Region 16.97 S 172.60 W
				H = 09 27 13.4 h = normal MB = 4.8
				D = 146.24 Az = 355.2 (NEIS)
				PKPV A 1.4s 14.0nm
17.	+eP	A	13 53 03	Hindu Kush Region 36.49 N 70.95 E
	epP	A	53 50	H = 13 45 13.8 h = 208 km MB = 5.3
	esP	A	54 11	D = 44.09 Az = 308.1 (NEIS)
				h = 219 km
				PV A 1.5s 78.0nm M = 5.0
17.	+eiP	AB	14 32 26	Iceland 64.67 N 21.23 W
	LmH	B	41.9	H = 14 27 32.0 h = normal MB=5.0 (NEIS)
	LmV	B	44.3	D = 22.14
				PV A 1.3s 118.0nm M = 5.2
				LmH B 14.5 0.9 /um 4.3
				LmV B 11 0.8 /um 4.5
17.	eP	AB	15 35 26.5	Peru 11.17 S 75.08 W
	epP	AB	35 54	H = 15 22 07.4 h = 111 km MB = 6.0
	esP	C	36 07	D = 96.46 Az = 39.8 (NEIS)
	eSKS	B	45 57	h = 108 km
	eiS	C	46 40	PV A 2.6s 173.0nm M = 6.1
	eSP	C	47 52	pPV A 2.2 283.5nm

Day	Phase		h m s	Remarks
cont. 17.	eSPP	C	15 48 40	LmH B 17.5s 0.8/um
	eSS	C	53.20	LmV B 18 1.3/um
	LmH	B	16 02.4	
	LmV	B	03.7	
17.	eP	AB	17 24 25	<u>Southwestern Ryukyu Islands</u>
	eS	B	34 51	25.14 N 125.56 E
	ePS	C	35 50	H = 17 11 50.8 h = 18 km MB=5.8 MS=5.6
	eSSS	C	45 00	D = 84.74 Az = 324.2 (NEIS)
	LmH	B	18 05.8	PV A 1.8s 155.0nm M = 5.9
	LmV	B	07.2	LmH B 14.5 9.7/um 6.3
				LmV B 16 11.2/um 6.4
17.	e	A	19 53 43	<u>Iran</u> 31.3 N 51.3 E
				H = 19 46 18 (BCIS)
				D = 35.11
17.	+eP	AB	21 08 35.5	<u>Java</u> 6.51 S 106.84 E
	epP	AB	09 09.5	H = 20 55 11.2 h = 131 km MB = 6.0
	ePP	B	12 35	D = 98.31 Az = 320.2 (NEIS)
	eSKS	B	18 58	h = 130 km
	eS	B	19 48	PV A 1.7s 51.5nm M = 5.8
	esS	B	20 48	SH B 10 1.8/um 6.3
	LmH	B	22 03.3	LmH B 14.5 1.2/um
	LmV	B	03.3	LmV B 16 1.2/um
17.	ePKIKP	A	23 25 40	<u>South of Fiji Islands</u>
	ePKHKP	A	25 48.5	26.12 S 179.54 E
	ePKP2	A	26 03	H = 23 06 44.1 h = 494 km MB = 5.3
	epPKHKP	A	27 41.5	D = 153.82 Az = 342.4 (NEIS)
				h = 487 km
				PKHKPV 1,4s 14.0nm
18.	eP	A	02 44 01	<u>Northern India</u> 26.95 N 71.70 E
				H = 02 34 55.3 h = 0 km MB = 5.0
				D = 50.88 Az = 314.7 (NEIS)
				Underground explosion (UPP)

Day	Phase		h m s	Remarks
18.	eP	A	23 44 49	Iceland 64.69 N 21.16 W
	eS	C	49 00	H = 23 39 56 h = 32 km MB = 4.7
	LmH	C	54.4	D = 22.12 Az = 114 (ISC)
	LmV	C	54.4	PV A 1.4s 32.6nm M = 4.6
				LmH C 13.5 0.5 /um 4.1
				LmV C 14.5 0.6 /um 4.3
19.	eP	A	13 21 27	Northern Italy 45.63 N 10.28 E
	ePg	A	21 51	H = 13 20 17.3 h = normal (NEIS)
	eSn	A	22 27	D = 5.10
	eSg	A	22 56	
19.	eP1	AC	22 05 21	Crete 35.49 N 26.33 E
	eiP2	A	05 24	H = 22 01 09.4 h = 83 km MB = 4.9
	eS	C	08 45	D = 18.53 Az = 329.4 (NEIS)
	LmV	B	14.1	P2V A 1.2s 44.6nm M = 4.6
	LmH	B	14.3	LmH B 12 0.5 /um
				LmV B 16 0.7 /um
20.	e	A	00 26 13	Off East Coast of Kamchatka
				51.37 N 159.70 E
				H = 00 14 19.7 h = normal MB = 4.6
				D = 74.81 Az = 339.6 (NEIS)
20.	ePn	A	04 20 07.5	Federal Republic of Germany
	eSn	A	20 40	49.83 N 7.70 E
	eSg	A	20 50	H = 04 19 25.2 h = 25 km
				D = 2.64 Az = 70.6 (NEIS)
20.	eP	A	10 50 49	Eastern Gulf of Aden 13.38 N 50.23 E
				H = 10 42 05.9 h = normal MB = 5.1
				D = 48.59 Az = 328.0 (NEIS)
				PV A 1.7s 27.2nm M = 5.0
20.	eP	A	17 47 50	Southwestern Kashmir 34.58 N 74.28 E
				H = 17 39 17.9 h = normal MB = 4.9
				D = 47.40 Az = 309.8 (NEIS)
				PV A 1.4s 11.6nm M = 4.7

Day	Phase		h m s	Remarks
21.	ePKIKP	A	05 06 21	Fiji Islands Region 21.21 S 178.74 W
	+iPKHKP	A	06 27	H = 04 47 40.7 h = 571 km MB = 5.2
	+iPKP2	A	06 34.5	D = 149.52 Az = 347.0 (NEIS)
21.	ePn	A	07 43 33.5	Switzerland 47.60 N 7.59 E
	i	A	43 34.2	H = 07 42 34.0 h = 5 km
	iPg	A	43 55	D = 4.04 Az = 39 (ISC)
	iSn	A	44 19.5	
	iSg	A	44 41	
21.	LmH	C	08 56.0	Java 8.61 S 111.21 E
	LmV	C	09 04.0	H = 07 57 10.0 h = 76 km MB = 5.0 (ISC)
				D = 102.7
				LmH C 30s 0.3 μm
				LmV C 20 0.3 μm
21.	ePKP	A	10 12 36	Fiji Region 16.4 S 178.5 W
				H = 09 53 01.1 h = 0 km
				D = 144.90 Az = 349 (ISC)
21.	LmH	C	12 25.0	LmH C 18s 0.3 μm
	LmV	C	25.0	LmV C 17 0.5 μm
21.	LmH	C	16 20.0	LmV C 18s 0.3 μm
	LmV	C	21.0	
22.	eP	A	07 43 39	Southern Greece 37.62 N 21.55 E
				H = 07 40 15.2 h = 129 km MB=3.9 (NEIS)
				D = 14.82
				PV A 1.2s 10.2nm M = 4.0
22.	ePKP	A	09 33 35	Tonga Islands 17.40 S 175.22 W
	e	A	33 40.5	H = 09 13 53.0 h = 21 km MB = 4.7
				D = 146.39 Az = 352.1 (NEIS)
				traces

Day	Phase		h m s	Remarks
22.	e	A	15 08 10	Loyalty Islands Region 22.59 S 172.26 E H = 14 48 06.7 h = 71 km MB = 5.2 D = 148.17 Az = 336.4 (NEIS)
23.	ePKP2 LmH LmV	A C C	07 03 55 08 15.5 15.5	North of New Zealand 29.84 S 177.07 W H = 06 43 29.6 h = 63 km MB = 4.7 (NEIS) D = 158.22 PKP2V A 1.2s 14.2nm LmH C 22 0.4/um LmV C 21 0.8/um
23.	eP1 eP2	A A	11 17 03 17 09	North Atlantic Ridge 27.32 N 44.37 W H = 11 08 24.8 h = normal MB=5.1 MS=5.0 D = 48.02 Az = 45.2 (NEIS) P2V A 1.4s 32.6nm M = 5.2
23.	eP	A	13 50 49.5	Southern Nevada 37.06 N 116.07 W H = 13 38 30.2 h = 5 km MB = 4.8 D = 81.29 Az = 30.6 (NEIS) traces
23.	ePKP	A	17 42 53	Fiji Islands Region 18.05 S 177.79 E H = 17 24 02.1 h = 450 km MB = 4.7 D = 146.62 Az = 349.1 (NEIS)
23.	eP e	A A	19 53 28 55 50	Yugoslavia 43.38 N 17.07 E H = 19 51 30.0 h = normal MB = 4.7 (NEIS) D = 8.17 PV A 1.0s 35.4nm M = 5.4
24.	eP	A	01 33 36.5	Near Coast of Libya 31.4 N 14.4 E H = 01 29 13 h = 0 km MB = 4.4 D = 19.31 Az = 355 (ISC)
24.	eP	A	09 34 04	Off East Coast of Honshu, Japan 33.02 N 141.08 E H = 09 21 25.6 h = 17 km MB = 4.6 (NEIS) D = 85.51 traces

Day	Phase		h m s	Remarks
24.	eP	A	20 38 04.5	Near East Coast of Kamchatka 53.08 N 159.81 E H = 20 26 34.9 h = normal MB = 5.0 D = 73.23 Az = 339.5 (NEIS) PV A 1.2s 28.4nm M = 5.1
24.	eP	A	21 31 26	Turkey 36.73 N 29.22 E H = 21 27 08.6 h = 37 km D = 18.79 Az = 323 (ISC) PV A 1.3s 10.9nm M = 3.9
25.	ePKHKP	A	08 27 30.5	Kermadec Islands Region 28.1 S 175.4 W H = 08 07 18 h = 0 km D = 156.83 Az = 349 (ISC)
25.	eP	A	09 43 46	Ryukyu Islands 27.57 N 129.63 E H = 09 31 13.5 h = 52 km MB = 4.8 D = 84.86 Az = 326 (ISC)
25.	eP	A	20 18 54	Eastern Greenland 70.97 N 20.94 W H = 20 13 30.0 h = normal MB = 4.7 D = 25.24 Az = 126.5 (NEIS) PV A 1.3s 26.2nm M = 4.7
26.	ePKP	AB	01 51 40	New Hebrides Islands 17.70 S 167.75 E
	e	AB	51 44.5	H = 01 32 11.2 h = 13 km MB = 5.8 MS=6.0
	ePP	B	54 52	D = 142.00 Az = 335.3 (NEIS)
	eSS	B	13 24	PKPV A 2.0s 21.4nm
	LmH	B	02 48.9	LmH B 18 2.7 µm M = 6.0
	LmV	B	03 02.6	LmV B 18 2.2 µm 5.9
26.	ePKIKP	A	06 06 17.5	Cook Islands Region 20.76 S 178.50 W
	ePKHKP	AB	06 22.5	H = 05 47 36.9 h = 565 km MB=5.7 (NEIS)
	ePKP2	A	06 29	D = 149.12
				PKIKPV traces
				PKHKPV A 1.2s 103.7nm
				PKP2V A 1.3 91.7nm

Day	Phase		h m s	Remarks
26.	ePKP	A	07 35 11	<u>New Hebrides</u> 17.67 S 167.85 E
				H = 07 15 44.4 h = 27 km MB = 4.7 (NEIS)
				D = 142.02
				PKPV A 1.7s 18.2nm
26.	ePKP	A	11 43 42.5	<u>New Hebrides Islands</u> 17.72 S 167.74 E
				H = 11 24 14.0 h = 19 km MB = 4.6
				D = 142.02 Az = 335.2 (NEIS)
26.	eP	A	13 10 27	<u>Southern Greece</u> 37.52 N 21.13 E
	LmH	B	15.8	H = 13 06 53.3 h = 57 km MB = 4.2 (NEIS)
	LmV	B	17.1	D = 14.78
				LmH B 14s 1.6 μm M = 4.3
				LmV B 12 0.7 μm 4.3
27.	eP	AB	04 53 00.5	<u>Kurile Islands</u> 50.83 N 157.35 E
	eS	B	05 02 32	H = 04 41 23.6 h = 47 km MB=5.6 MS=5.5
	LmH	B	29.5	D = 74.78 Az = 338.2 (NEIS)
	LmV	B	29.6	PV A 2.4s 166.0nm M = 5.6
				PV B 10 0.86 μm 5.7
				LmH B 20 5.3 μm 5.8
				LmV B 20 5.2 μm 5.9
27.	eP	A	05 14 03	<u>Guerrero, Mexico</u> 17.30 N 98.96 W
	epP	A	14 13.5	H = 05 01 11.3 h = 58 km MB = 5.3
	ePP	A	17 45	D = 89.16 Az = 36.6 (NEIS)
				h = 70 km
				PV A 1.9s 45.5nm M = 5.5
27.	eP	A	10 50 35	<u>Mindanao, Philippine Islands</u>
	e	A	50 41	8.53 N 123.16 E
	e	A	50 51	H = 10 37 05.8 h = 35 km MB = 5.2
	ePP	A	54 35	D = 96.76 Az = 323.4 (NEIS)
	LmH	B	11 39.6	PV A 1.9s 19.0nm M = 5.3
	LmV	B	39.6	LmH B 20 1.3 μm 5.4
				LmV B 20 1.6 μm 5.5

Day	Phase		h m s	Remarks
27.	eSg	A	11 25 30	Austria 47.8 N 16.2 E H = 11 23 21 (BCIS) D = 4.10
27.	eP	A	12 28 17	North Atlantic Ridge 15.7 N 46.7 W H = 12 18 24.5 h = 33 km MB = 4.9 (NEIS) D = 58.05
27.	eP ipP LmH LmV	AB AB B B	14 12 42 12 47 46.7 46.9	Eastern Alaska to Vancouver Islands 60.33 N 146.02 W H = 14 01 43.5 h = 21 km MB = 5.5 (NEIS) D = 67.92 h = 19 km PV A 1.0s 63.0nm M = 5.7 LmH B 16 0.6/um 4.9 LmV B 16 0.6/um 5.0
28.	ePKP2	A	03 15 37	Kermadec Islands Region 31.81 S 179.36 E H = 02 55 50.9 h = 450 km MB = 5.0 (NEIS) D = 159.10 PKP2V A 1.4s 55.8nm
29.	eP	A	04 27 15	Carlsberg Ridge 6.31 N 60.65 E H = 04 17 08.9 h = normal MB = 5.1 (NEIS) D = 60.06 PV A 1.8s 20.3nm M = 4.9
29.	e	A	19 27 06	
30.	eP epP	A A	01 11 35.5 11 43	Vancouver Island Region 49.06 N 128.39 W H = 00 59 56.1 h = normal MB=4.8 MS=4.2 D = 74.93 Az = 25.1 (NEIS) h = 28 km PV A 1.2s 12.2nm M = 4.8
30.	eP	A	01 16 58	Kurile Islands 47.98 N 154.53 E H = 01 05 09.0 h = normal MB = 4.6 (NEIS) D = 76.71

Day	Phase		h m s	Remarks
30.	eSb	A	01 29 04	Poland 50.33 N 18.83 E H = 01 26 32.5 M = 2.5 (WAR) D = 4.7
30.	e	A	16 55 12	
31.	eP	A	02 08 19	Near East Coast of Kamchatka 53.96 N 159.18 E H = 01 57 03.2 h = 111 km MB = 4.9 (NEIS) D = 72.27
31.	eP LmH LmV	A C C	03 24 56.5 04 02.2 02.3	Unimak Island Region 53.60 N 163.82 W H = 03 13 10.7 h = 33 km MB = 4.8 (NEIS) D = 76.01 PV A 1.5s 35.2nm M = 5.2 LmH C 20 0.5 μm 4.8 LmV C 21 0.8 μm 5.1
31.	+iP ePn ePP	A A A	03 34 49.0 36 20 36 26	Eastern Kazakh SSR 49.95 N 78.84 E H = 03 26 57.4 h = 0 km MB = 5.9 D = 41.60 Az = 297.8 (NEIS) Underground explosion MB = 6.9 (UPP) PV A 1.1s 226.0nm M = 5.8 PH A 1.1 136.5nm 5.8 PnV A 0.8 28.8nm PPV A 1.4 55.8nm 5.2
31.	ePKHKP	A	06 28 21.5	West of Tonga 20.35 S 177.95 W H = 06 09 35.6 h = 573 km MB = 4.3 (NEIS) D = 148.83
31.	e	A	08 00 07	Tonga Region 17.3 S 172.4 W H = 07 40 19 h = 53 km D = 146.60 Az = 355 (ISC)
31.	iPn iSn	A A	08 00 32.5 00 57	Czechoslovakia 51.00 N 14.41 E Explosion H = 08 00.0 yield 13 t (PRU) D = 1.79

Day	Phase		h m s	Remarks
31.	eP	A	09 22 08	Near East Ciast of Kamchatka
				52.96 N 160.09 E
				H = 09 10 37.9 h = normal MB = 5.1 (NEIS)
				D = 73.40
				PV A 1.4s 51.1nm M = 5.2
31.	eP	A	11 24 55	Atlantic-Indian Rise 28.10 S 63.44 E
	epP	A	25 04	H = 11 11 54.3 h = normal MB = 5.4
				D = 90.78 Az = 330.0 (NEIS)
				h = 32 km
				pPV A 1.4s 27.9nm
31.	eP	A	12 33 55.5	Iceland 65.05 N 20.9 W
				H = 12 29 00.1 h = 37 km MB = 4.4
				D = 22.18 Az = 115 (ISC)
31.	eP	A	14 17 51	Northwestern Mexico 27.23 N 111.24 W
	ePP	A	21(20)	H = 14 04 59.9 h = normal MB = 5.3 (NEIS)
	eSKS	C	28 25	D = 87.52
	eS	C	28 36	PV A 1.7s 33.4nm M = 5.3
	eSS	C	34 20	LmH B 18 15.9 μm 6.5
	LmV	B	58.2	LmV B 19.5 19.3 μm 6.5
	LmH	B	58.4	

Day	Phase		h m s	Remarks
1.	eP	A	16 10 01	Southern Greece 36.18 N 22.49 E H = 16 07 08.4 h = 39 km MB = 4.5 D = 16.45 Az = 334.9 (NEIS)
1.	eP	A	19 11 41	Near East Coast of Kamchatka 53.55 N 160.37 E H = 19 00 13.9 h = 40 km MB = 4.5 D = 72.90 Az = 339.8 (NEIS)
1.	eP LmH LmV	A B B	20 28 14 39.8 41.5	North Atlantic Ocean 55.29 N 35.28 W H = 20 22 23.0 h = normal MB = 4.7 D = 28.19 Az = 79.7 (NEIS) PV A 1.7s 33.4nm M = 4.8 LmH B 15.5 0.6 μm 4.3 LmV B 14 0.6 μm 4.4
2.	eP	A	04 26 48	Ryukyu Islands 29.24 N 130.50 E H = 04 14 19.8 h = normal MB = 4.7 D = 83.92 Az = 325.9 (NEIS) traces
2.	ePn ePg eSg	A A A	05 27 50 28 25 30 18	Adriatic Sea 42 1/4 N 16 1/4 E H = 05 25 34 (BCIS) D = 9.0
2.	eP	A	07 03 11	Mid-Indian Rise 23.96 S 70.06 E H = 06 50 12.3 h = normal MB = 5.1 D = 90.38 Az = 327.1 (NEIS) PV A 1.8s 20.3nm M = 5.1
2.	+ePKP	A	12 37 04	Tonga Islands 16.64 S 172.99 W H = 12 17 22.9 h = 7 km MB = 5.3 MS=5.1 D = 145.88 Az = 354.8 (NEIS) PKPV A 1.6s 115.4nm
2.	LmH	B	16 55.1	Talaud Islands 2.68 N 125.28 E H = 15 59 51.4 h = normal

Day	Phase		h m s	Remarks
cont. 2.	LmV	B	17 06.8	MB = 5.4 MS = 5.5 (NEIS) D = 102.6 LmH B 19s 1.9/um M = 5.7 LmV B 19 1.9/um 5.7
2.	ePKHKP ePKP2	A A	22 47 01 47 11	South of Fiji Islands 22.91 S 179.79 W H = 22 28 11.5 h = 577 km MB = 4.8 D = 150.92 Az = 345.0 (NEIS)
2.	eP	A	23 16 16	Colombia 5.34 N 76.88 W H = 23 03 46.2 h = 64 km MB = 5.2 D = 84.95 Az = 39.7 (NEIS)
3.	e	A	01 27 13	
3.	+ePKP LmH LmV	A C C	06 34 12.5 07 36.5 41.6	Tonga Islands 15.41 S 173.33 W H = 06 14 38.5 h = normal MB=5.4 MS=5.0 D = 144.63 Az = 354.6 (NEIS) PKPV A 1.8s 50.7nm
3.	eP ePP e(PcP)	A A A	11 53 36.5 55 14.5 55 25	Afghanistan-USSR Border Region 36.86 N 71.40 E H = 11 45 36.2 h = 100 km MB = 5.3 D = 44.15 Az = 307.8 (NEIS) PV A 1.5s 20.1nm M = 4.7
3.	eP ePP	A A	23 43 19.5 45 06	Tadzhik SSR 39.13 N 71.44 E H = 23 35 25.8 h = 60 km MB = 5.0 D = 42.81 Az = 305.9 (NEIS) PV A 1.3s 10.9nm M = 4.5
4.	ePKP	A	02 38 40	South of Fiji Islands 21 S 177 E H = 02 18 59 MB = 4.1 (NORSAR) D = 148.4
4.	-iPKP epPKP esPKP	AB B B	04 33 21 34 30 35 00	Tonga Islands 15.85 S 175.10 W H = 04 14 15.9 h = 276 km MB = 6.0 D = 144.87 Az = 352.6 (NEIS)

Day	Phase		h m s	Remarks
cont. 4.	-iPP	B	04 36 41	h = 281 km PKPV A 1.6s 1700.0nm PKPV B 13 9.2/um PPV B 7 4.6/um M = 6.7
4.	eP	A	14 22 09.5	Southern Italy 38.9 N 17.8 E
	LmV	C	26.7	H = 14 19 08 (BCIS)
	LmH	C	29.6	D = 12.64 LmH C 16s 1.1/um M = 4.0 LmV C 22 0.4/um
4.	eP	A	15 24 06	North Atlantic Ridge 10.84 N 42.55 W
	LmV	B	47.2	H = 15 14 03.4 h = normal MB=5.0 MS=5.0
	LmH	B	47.9	D = 59.32 Az = 36.9 (NEIS) PV A 1.5s 22.6nm M = 5.0 LmH B 17 0.6/um 4.8 LmV B 20 0.7/um 4.9
4.	ePg	A	17 45 24.5	
	eSg	A	45 45.5	D = 1.7
4.	e	A	19 38 27	
4.	e	A	24 00 40	Samoa Region 16.2 S 172.8 W H = 23 40 43.2 h = 33 km D = 145.42 Az = 355 (ISC)
5.	eP	A	00 13 03	Szechwan Province, China
	LmH	B	42.9	29.38 N 99.46 E
	LmV	B	45.7	H = 00 02 10.8 h = normal MB = 5.1 D = 66.59 Az = 316.1 (NEIS) PV A 1.6s 16.5nm M = 4.9 LmH B 19 0.8/um 5.0 LmV B 15 0.6/um 5.0
5.	e	A	01 15 14	

Day	Phase		h m s	Remarks
5.	ePKIKP	A	02 05 58	New Britain Region 6.59 S 149.16 E H = 01 47 03.6 h = 49 km MB = 5.0 D = 123.68 Az = 328.9 (NEIS)
5.	ePKIKP	A	05 39 26	New Hebrides Islands 14.64 S 167.34 E H = 05 20 16.7 h = 159 km MB = 5.0 (NEIS) D = 139.0
5.	ePKHKP	A	08 37 34.5	Fiji Islands Region 20.69 S 177.74 W H = 08 18 24.9 h = 345 km MB = 4.8 D = 149.20 Az = 348.3 (NEIS) PKHKPV A 1.0s 17.7nm
5.	iPn iSg	A A	11 20 52 21 13	Czechoslovakia 50.59 N 14.05 E Explosion yield 30.2 t (PRU) D = 1.53
5.	LmH	C	13 08.7	Near East Coast of Honshu 37.00 N 141.78 E H = 12 25 51.8 h = 50 km MB = 4.5 (ISC) D = 82.3 LmH C 24s 0.3/um M = 4.5
5.	ePKP e	A A	22 20 25 20 35	Tonga Islands 15.04 S 173.79 W H = 22 00 49.2 h = normal MB = 5.0 D = 144.22 Az = 354.1 (NEIS) PKPV A traces
6.	LmH LmV	B C	13 10.5 15.5	North Eastern China 37.54 N 115.10 E H = 12 30 59 h = 45 km MB = 4.7 (ISC) D = 69.5 LmH B 15s 1.1/um M = 5.1 LmV C 16 2.8/um 5.6
6.	eP epP	A A	17 13 41 13 52	Kurile Islands 45.00 N 150.76 E H = 17 01 41.9 h = normal MB = 4.9 D = 78.36 Az = 334.8 (NEIS) h = 41 km PV A 1.5s 27.7nm M = 5.1

Day	Phase		h m s	Remarks
6.	ePKIKP	A	17 29 45	Fiji Islands Region 17.83 S 178.58 W
	ePKHKP	A	29 47	H = 17 11 09.0 h = 572 km MB = 4.9
				D = 146.26 Az = 348.3 (NEIS)
				traces
6.	ePS	C	18 45 40	New Ireland Region 2.88 S 149.09 E
	ePPS	C	47 28	H = 18 15 33.4 h = 37 km MB=5.3 MS=5.7
	eSS	C	52 33	D = 120.47 Az = 330.0 (NEIS)
	LmV	B	19 27.0	LmH B 21s 3.2/um M = 5.9
	LmH	B	27.2	LmV B 21 4.3/um 6.1
6.	eP	A	18 47 47	Carlsberg Ridge 4.24 N 66.54 E
				H = 18 37 07.7 h = normal MB=5.4 MS=5.5
				D = 65.07 Az = 324.9 (NEIS)
6.	+eP	A	19 11 14	Hindu Kush Region 36.38 N 70.73 E
	e	A	13 23	H = 19 03 26.0 h = 214 km MB = 5.3
				D = 44.02 Az = 308.2 (NEIS)
				PV A 1.3s 52.5nm M = 4.8
6.	ePKP	A	20 20 18.5	Tonga Islands 15.24 S 173.47 W
				H = 20 00 45.0 h = normal MB=5.2 MS=5.2
				D = 144.44 Az = 354.4 (NEIS)
				PKPV A 1.3s 10.9nm
6.	ePKHKP	A	24 04 45.5	Tonga Region 22.2 S 174.2 W
				H = 23 44 46 h = 0 km
				D = 151.21 Az = 352 (ISC)
7.	ePKHKP	A	00 41 18	Tonga Region 22.4 S 174.3 W
	ePKP2	A	41 29	H = 00 21 24 h = 33 km
				D = 151.40 Az = 352 (ISC)
7.	ePKP	AB	07 07 11	Tonga Islands 15.44 S 175.32 W
	LmV	B	08 08.5	H = 06 47 36.3 h = normal MB=5.2 MS=6.0
	LmH	B	09.2	D = 144.44 Az = 352.4 (NEIS)
				PKPV A 2.0s 51.3nm
				LmH B 22 2.5/um M = 5.9
				LmV B 23 3.7/um 6.1

Day	Phase		h m s	Remarks
7.	e(Pg)	A	13 40 57	
	e	A	41 14.5	
	e	A	41 33	
	e	A	42 21.5	
7.	e(P)	A	14 49 46	Greece-Albania Border Region
	LmH	B	54.3	39.07 N 20.40 E
	LmV	B	55.9	H = 14 46 30.3 h = 35 km MB = 4.2
				D = 13.14 Az = 334.7 (NEIS)
				LmH B 14s 1.4 µm M = 4.2
				LmV B 13 0.8 µm
7.	eP	A	18 02 10.5	Aleutian Islands Region 50.93 N 170.63 W
				H = 17 50 08.8 h = normal MB = 5.0
				D = 78.79 Az = 358.5 (NEIS)
				PV A 1.1s 28.2nm M = 5.2
7.	ePKHKP	A	22 47 52	South of Fiji Islands 24.29 S 177.53 W
				H = 22 28 26.1 h = 280 km MB = 4.6
				D = 152.74 Az = 342.9 (NEIS)
7.	eP	AC	23 01 42	South of Panama 5.66 N 82.58 W
	e	C	02 31	H = 22 48 48.5 h = normal MB=5.4 MS=5.9
	eSKS	C	12 08	D = 88.33 Az = 39.4 (NEIS)
	eS	C	12 28	PV A 2.2s 65.5nm M = 5.6
	ePS	C	13 20	LmH B 22 5.2 µm 5.9
	eSS	C	18 20	LmV B 21.5 5.2 µm 5.9
	LmH	B	35.5	
	LmV	B	35.6	
8.	ePKHKP	A	12 55 15.5	South of Fiji Islands 24.49 S 179.57 W
	ePKP2	A	55 26.5	H = 12 36 15.2 h = 518 km MB = 5.1
				D = 152.49 Az = 344.5 (NEIS)
8.	e(PS)	C	17 46 30	Solomon Islands 7.23 S 155.15 E
	e(PPS)	C	48 00	H = 17 15 25.1 h = 33 km MB=5.1 MS=5.3
	e(SS)	C	53 32	D = 127.19 Az = 331.6 (NEIS)
	LmH	B	18 35.7	LmH B 18s 0.5 µm M = 5.1
	LmV	B	35.7	LmV B 18 0.5 µm 5.2

Day	Phase		h m s	Remarks
9.	eSg	A	00 21 18.5	France 45.9 N 6.4 E
				H = 00 18 05 (BCIS)
				D = 5.90
9.	ePKP	AC	03 21 11	Samoa Islands Region 16.54 S 172.69 W
	LmH	B	04 34.0	H = 03 01 33.4 h = normal MB=5.1 MS=5.1
	LmV	B	41.0	D = 145.81 Az = 355.1 (NEIS)
				PKPV A 2.2s 65.5nm
9.	e	A	10 54 54	Near Coast of Northern Peru
	eS	C	11 06 00	5.75 S 80.86 W
	eSS	C	12 25	H = 10 41 22.1 h = 52 km MB = 5.1
	e	C	21 00	D = 95.98 Az = 39.8 (NEIS)
	e	C	22 00	PV A 2.0s 30.0nm M = 5.5
	LmH	B	50.0	LmH B 16 0.35 μm
	LmV	B	50.0	LmV B 17 0.4 μm
9.	eP	AB	14 29 27	Near Coast of Northern Peru
	e	A	29 36.5	5.75 S 80.98 W
	e	A	30 23	H = 14 16 03.7 h = 50 km MB = 5.7
	ePP	A	33 20	D = 96.05 Az = 39.8 (NEIS)
	eSKS	C	40 10	PV A 1.8s 40.5nm M = 5.7
	eS	C	40 40	LmH B 17 0.8 μm 6.3
	ePS	C	42 10	LmV B 17 0.9 μm 6.3
	eSS	C	47 00	
	LmH	B	15 17.5	
	LmV	B	19.0	
10.	eP	A	04 28 05	Morocco 33.61 N 3.79 W
				H = 04 23 28.6 h = 39 km MB = 4.3
				D = 20.43 Az = 29.0 (NEIS)
10.	eP	A	07 47 19	Off Coast of Northern California
				40.45 N 125.08 W
				H = 07 35 00.5 h = 22 km MB = 4.8
				D = 81.68 Az = 26.2 (NEIS)

Day	Phase		h m s	Remarks
10.	eP	A	09 08 35.5	Kurile Islands 46.79 N 152.79 E H = 08 56 43.0 h = normal MB = 4.6 D = 77.32 Az = 335.9 (NEIS) PV A 1.0s 7.9nm M = 4.7
10.	eP	A	10 26 41	North Atlantic Ridge 44.73 N 28.31 W H = 10 20 58.2 h = normal MB = 4.3 D = 27.27 Az = 63.1 (NEIS) traces
10.	ePg	A	11 52 52	Czechoslovakia 50.3 N 14.4 E
	eSg	A	53 13.5	H = 11 52 22 h = 0 km D = 1.78 Az = 281 (ISC)
10.	eP	A	14 46 04	Alaska Peninsula 54.77 N 161.65 W H = 14 34 22.3 h = 11 km MB = 4.8 D = 74.81 Az = 4.4 (NEIS) PV A 1.2s 18.6nm M = 5.0
10.	e(sPKP)	A	19 23 20.5	New Hebrides 15.82 S 167.91 E
	e(PP)	A	25 28	H = 19 02 58.6 h = 182 km MB = 5.3 (ISC) D = 140.3
11.	eP	A	02 11 09	Eastern Mediterranean Sea 34.63 N 28.51 E H = 02 06 33.9 h = 43 km MB = 4.1 D = 20.18 Az = 327.6 (NEIS)
11.	e	A	18 04 59	traces
11.	eP	A	20 32 37	Andreanof Islands, Aleutian Is. 51.82 N 173.53 W H = 20 20 45.0 h = 58 km MB = 4.8 D = 77.72 Az = 356.6 (NEIS) PV A 0.8s 7.7nm
11.	ePKP2	A	22 35 45	Kermadec Islands 29.92 S 178.46 W
	e	A	35 48	H = 22 15 12.5 h = 16 km MB = 4.7 (NEIS)
	e	A	36 06.5	D = 157.9

Day	Phase		h m s	Remarks
cont. 11.	LmH	B	23 58.3	PKP2V A 1.4s 16.3nm
	LmV	B	58.5	LmH B 20 0.8 um M = 5.4
				LmV B 19 0.6 um 5.4
12.	eP	A	10 25 16	<u>Jordan-Syria Region</u> 34.04 N 37.69 E
	e	A	25 36	H = 10 19 51.1 h = normal MB = 4.6
				D = 25.24 Az = 319.0 (NEIS)
				PV A 1.6s 44.0nm M = 4.8
12.	LmH	B	14 43.2	<u>Near Coast of Guerrero, Mexico</u>
	LmV	B	43.4	16.77 N 99.26 W
				H = 13 46 40.8 h = 43 km MB = 4.9 (ISC)
				D = 89.8
				LmH B 16s 0.3 um M = 4.6
				LmV B 14 0.3 um 4.8
12.	eP	A	16 13 55	<u>Iceland</u> 64.88 N 20.82 W
	LmH	B	23.4	H = 16 08 58.7 h = 16 km MB = 4.8
	LmV	B	25.7	D = 22.07 Az = 114.7 (NEIS)
				PV A 1.6s 44.0nm M = 4.6
				LmH B 14.5 0.7 um 4.2
				LmV B 11.5 0.7 um 4.5
12.	eP	AB	16 37 12.5	<u>Near Coast of Venezuela</u>
	+i	A	37 14.5	10.56 N 63.38 W
	eS	B	46 39	H = 16 25 47.6 h = 34 km MB=5.7 MS=6.1
	eSKS	C	47 14	D = 72.39 Az = 40.2 (NEIS)
	eiSS	C	51 20	PV A 3.0s 579.0nm M = 6.1
	LmH	B	17 05.5	PV B 8 1.20 um 6.0
	LmV	B	05.6	LmH B 20.5 7.8 um 6.0
				LmV B 21 8.1 um 6.0
12.	eP	A	16 58 25.5	<u>Fox Islands, Aleutian Is.</u>
				52.44 N 170.20 W
				H = 16 46 34.4 h = 46 km MB = 5.2
				D = 77.28 Az = 358.8 (NEIS)
				PV A 1.1s 24.2nm M = 5.1

Day	Phase		h m s	Remarks
12.	e(P)	A	17 19 52	South Burma 15.4 N 95.0 E
				H = 17 07 52 h = 33 km MB = 4.2
				D = 74.69 Az = 319 (ISC)
12.	eP	AB	18 00 04.5	Iceland 64.77 N 21.05 W
	eS	C	04 15	H = 17 55 08.7 h = 13 km MB=5.5 MS=5.3
	LmH	B	09.4	D = 22.11 Az = 114.1 (NEIS)
	LmV	B	09.7	PV A 1.5s 262.0nm M = 5.5
				SH C 15 12.5/um 6.0
				LmH B 13.7 8.1/um 5.3
				LmV B 15 7.2/um 5.4
12.	ePKP	A	22 24 42	Fiji Islands Region 17.82 S 178.43 W
				H = 22 06 00.0 h = 555 km MB = 4.8
				D = 146.28 Az = 348.5 (NEIS)
13.	ePP	A	04 24 26	Strait of Gibraltar 36.88 N 4.08 W
				H = 04 20 06.3 h = 51 km MB = 4.1
				D = 17.78 Az = 34.3 (NEIS)
				PPV A 1.6s 13.7nm
13.	eP	A	07 15 57	Turkey 37.7 N 32.8 E
				H = 07 11 23 h = 0 km
				D = 19.85 Az = 317 (ISC)
13.	eP	A	13 45 12	Iceland 64.97 N 22.0 W
				H = 13 40 08.1 h = 0 km MB = 4.3
				D = 22.57 Az = 113 (ISC)
				PV A 1.2s 8.1nm
14.	ePn	A	11 41 12	Czechoslovakia 50.58 N 14.0 E
	iSg	A	41 32.8	Explosion yield 22.5 t
				H = 11 40.7 (PRU)
				D = 1.52
14.	ePKP	A	13 43 32.5	Scotia Sea 60.65 S 37.60 W
				H = 13 24 40.9 h = normal MB=5.4 MS=5.6
				D = 117.79 Az = 33.0 (NEIS)
				PKPV A 1.0s 11.8nm

Day	Phase		h m s	Remarks
14.	ePKHKP	A	18 58 03.5	South of Fiji Islands 22.09 S 179.69 W H = 18 39 18.8 h = 603 km MB = 5.1 D = 150.15 Az = 345.5 (NEIS)
15.	eP	A	00 57 19	Eastern Caucasus 42.94 N 45.21 E
	e	A	01 02 08	H = 00 52 05.9 h = 45 km MB = 4.7
	ePcS	B	04 30	D = 24.09 Az = 300.4 (NEIS)
	LmH	B	08.2	PV A 2.0s 68.5nm M = 4.8
	LmV	B	08.9	LmH B 15 0.55 µm 4.2
				LmV B 14 0.55 µm 4.3
15.	e	A	02 49 13.5	Rat Islands, Aleutian Is.
	epP	A	49 25.5	52.26 N 178.79 E
	e	AB	49 32	H = 02 37 13.8 h = 157 km MB = 5.7
				D = 76.90 Az = 351.7 (NEIS)
				PV A 1.7s 84.8nm M = 5.2
15.	eP	A	03 41 29	Eastern Gulf of Aden 13.72 N 50.47 E H = 03 32 48.1 h = normal MB = 5.2 D = 48.43 Az = 327.7 (NEIS) PV A 2.0s 51.3nm M = 5.2
15.	+eP	AB	07 14 10	Szechwan Province, China
	eS	C	23 20	28.27 N 103.95 E
	LmH	B	41.5	H = 07 03 00.2 h = 39 km MB=5.5 MS=5.3
	LmV	B	46.8	D = 70.10 Az = 317.4 (NEIS)
				PV A 2.2s 65.5nm M = 5.3
				PV B 4.5 0.4 µm 5.7
				LmH B 21 2.0 µm 5.3
				LmV B 16 0.7 µm 5.1
15.	eP	A	10 30 12.5	Szechwan Province, China 31.76 N 100.09 E H = 10 19 29.2 h = normal MB = 4.9 D = 65.27 Az = 315.5 (NEIS) PV A 1.9s 22.8nm M = 4.9

Day	Phase		h m s	Remarks
15.	eP	A	14 38 46	Szechwan Province, China 28.33 N 104.00 E H = 14 27 29.3 h = 10 km MB = 5.2 D = 70.08 Az = 317.4 (NEIS) PV A 2.2s 32.7nm M = 5.1
15.	eSg	A	17 44 37	France 45.9 N 6.9 E H = 17 41 39 (BCIS) D = 5.74
16.	ePKHKP	A	13 37 22	Fiji Islands Region 19.66 S 178.54 W H = 13 18 46.0 h = 647 km MB = 4.4 D = 148.05 Az = 347.8 (NEIS) PKHKPV A 1.0s 11.81nm
17.	eP	A	02 12 36.5	Kurile Islands 48.42 N 154.12 E H = 02 00 50.0 h = normal MB = 4.6 D = 76.19 Az = 336.5 (NEIS) PV A 0.8s 11.5nm M = 4.9
17.	eP	A	02 30 16.5	Kurile Islands 48.23 N 154.35 E
	LmH	C	59.5	H = 02 18 29.5 h = normal MB=5.1 MS=4.3 D = 76.43 Az = 336.6 (NEIS) PV A 1.8s 33.8nm M = 5.1
17.	eP	A	06 09 14.5	Southern Sinkiang Province
	LmH	B	32.6	40.55 N 89.6 E
	LmV	B	32.7	H = 05 59 52.5 h = 0 km MB = 4.5 D = 53.13 Az = 309 (ISC) Probably atmospheric explosion (UPP) LmH B 13s 0.35 µm M = 4.6 LmV B 15 0.45 µm 4.7
17.	eP	A	07 29 57	Iran 33.66 N 57.01 E H = 07 22 49.1 h = 35 km MB = 4.8 D = 37.03 Az = 311.2 (NEIS)

Day	Phase		h m s	Remarks
17.	eP	A	16 48 36	Hokkaido, Japan Region 41.34 N 142.47 E
				H = 16 36 37.9 h = 60 km MB = 5.0
				D = 78.82 Az = 330.6 (NEIS)
				PV A 1.2s 12.2nm M = 4.7
17.	eP	A	19 28 51.5	Southern Italy 38.70 N 17.77 E
	LmH	B	33.6	H = 19 25 52.8 h = normal MB = 3.8
	LmV	B	36.4	D = 12.72 Az = 341.9 (NEIS)
				PV A 1.4s 9.3nm M = 4.5
				LmH B 16 0.5/um 3.6
				LmV B 15 0.3/um
17.	ePg	A	21 27 18	Austria 47.1 N 10.6 E
	eSg	A	28 03	H = 21 26 08 (BCIS)
				D = 3.59
18.	eP	A	08 29 25	Greece 38.54 N 20.42 E
	LmH	B	36.0	H = 08 26 12.9 h = normal MB=4.8 MS=4.4
	LmV	B	36.1	D = 13.62 Az = 335.6 (NEIS)
				PV A 0.8s 13.5nm M = 4.9
				LmH B 12.5 1.7/um 4.3
				LmV B 12 1.6/um
18.	ePKHKP	A	08 50 39	South of Fiji Islands 24.79 S 179.86 E
	ePKP2	A	50 51.5	H = 08 31 38.4 h = 506 km MB = 5.0
				D = 152.64 Az = 343.6 (NEIS)
19.	eP	A	03 08 30.5	Atlantic-Indian Rise 33.42 S 56.88 E
	X	A	08 45	H = 02 55 19.7 h = normal MB=5.0 MS=6.6
	LmH	B	46.0	D = 92.74 Az = 333.1 (NEIS)
	LmV	B	50.0	PV A traces
				XV A 1.5s 15.1nm
				LmH B 18 1.2/um M = 5.4
				LmV B 12.5 0.4/um 5.1
19.	eP	A	03 19 59	Eastern Siberia 63.16 N 150.85 E
				H = 03 09 38.0 h = normal MB=5.0 MS=4.8
				D = 62.09 Az = 331.9 (NEIS)
				PV A 0.9s 15.6nm M = 5.1

Day	Phase		h m s	Remarks
19.	eP	A	11 46 56	Unimak Island Region 53.82 N 163.43 W H = 11 35 11.7 h = normal MB = 4.7 D = 75.82 Az = 3.2 (NEIS) PV A 1.1s 16.1nm M = 5.0
19.	eP	A	16 12 18.5	Southern Nevada 37.20 N 116.19 W H = 15 59 59.9 h = 5 km MB = 5.0 D = 81.22 Az = 30.6 (NEIS) PV A 1.2s 12.2nm M = 4.8 Probably underground explosion
19.	eP	A	19 34 01	Off Coast of Northern California 41.93 N 126.78 W H = 19 21 48.9 h = normal MB=4.9 MS=4.2 D = 80.90 Az = 25.3 (NEIS) PV A 2.0s 21.4nm M = 4.8
20.	ePKIKP ePKHKP	A A	06 54 42.5 54 51	South of Fiji Islands 26.04 S 179.27 E H = 06 35 52.7 h = 540 km MB = 5.3 D = 153.67 Az = 342.1 (NEIS)
20.	-iPn eSn LmH LmV	A A B B	09 30 23.2 32 07 32.9 33.6	Yugoslavia 44.35 N 17.71 E H = 09 28 33.4 h = normal MB = 5.1 D = 7.53 Az = 328.9 (NEIS) PV A 1.0s 82.7nm M = 5.7 LmH B 11 1.7/um 4.0 LmV B 13 1.6/um
20.	iPn iPg eiSn eSg LmH	A A AB AB B	17 09 46.1 10 07 11 04 11 32 11.7	Yugoslavia 45.97 N 15.53 E H = 17 08 27.3 h = 47 km MB = 4.5 D = 5.36 Az = 332.3 (NEIS) LmH B 7s 16.6/um M = 4.9
20.	e	A	18 36 30	

Day	Phase		h m s	Remarks
20.	+ePn	AB	22 27 49.5	Yugoslavia 46.09 N 15.48 E
	eSn	A	28 46	H = 22 26 31.8 h = normal MB = 4.4
	eSg	A	29 10	D = 5.24 Az = 332.0 (NEIS)
	LmH	B	29.5	LmH B 6.8s 5.1/um M = 4.4
	LmV	B	29.5	LmV B 4.0 1.2/um
21.	eSg	A	01 03 05	Poland 50.28 N 18.85 E
				H = 01 00 33.9 M = 2.9 (WAR)
				D = 4.65
21.	eP	A	08 52 20	North Atlantic Ocean 57.84 N 32.63 W
	LmH	B	09 03.2	H = 08 46 45.0 h = normal MB=4.8 MS=4.4
	LmV	B	03.4	D = 26.43 Az = 86.5 (NEIS)
				PV A 1.5s 20.1nm M = 4.5
				LmH B 14.8 1.0/um 5.5
				LmV B 14.5 1.2/um 5.7
21.	eP	A	16 12 31	Western Gulf of Aden 12.81 N 46.84 E
				H = 16 03 57.2 h = normal MB = 4.5
				D = 47.38 Az = 330.1 (NEIS)
				PV A 1.3s 13.1nm M = 4.8
21.	eP	A	21 06 32.5	East of Lake Baikal 56.50 N 117.26 E
	LmH	B	33.3	H = 20 56 48.7 h = normal MB=5.3 MS=4.5
	LmV	B	36.3	D = 56.87 Az = 312.9 (NEIS)
				PV A 1.4s 41.9nm M = 5.3
				LmH B 13 0.6/um 4.9
				LmV B 11 0.4/um 4.9
22.	eSg	A	01 49 48	Yugoslavia 46.2 N 15.5 E
				H = 01 47 09 (BCIS)
				D = 5.15
22.	ePKIKP	A	07 30 33	Kermadec Islands Region
	ePKP2	A	31 00.5	27.66 S 178.02 W
				H = 07 10 58.3 h = 164 km MB = 5.2
				D = 155.90 Az = 344.9 (NEIS)
				PKIKPV A traces

Day	Phase		h m s	Remarks
22.	ePKIKP	AB	08 31 53.5	<u>Easter Island Region</u> 22.07 S 113.59 W
	ePP	AB	33 58	H = 08 12 47.5 h = normal MB=5.9 MS=5.7
	ePKS	B	35 18	D = 128.94 Az = 42.0 (NEIS)
	LmH	B	09 28.8	PKIKPV A 2.0s 128.0nm
	LmV	B	29.0	LmH B 17.5 1.2 μm M = 5.6
				LmV B 18 1.6 μm 5.8
22.	ePKHKP	A	10 19 42	<u>Tonga Islands</u> 20.80 S 174.45 W
				H = 09 59 53.0 h = normal MB=5.1 MS=4.9
				D = 149.83 Az = 352.3 (NEIS)
				PKHKPV A 1.1s 20.2nm
22.	eP	A	10 41 58	<u>Near East Coast of Honshu, Japan</u>
				39.97 N 142.81 E
				H = 10 29 50.8 h = 40 km MB = 5.2
				D = 80.15 Az = 330.9 (NEIS)
				PV A 1.3s 21.8nm M = 5.0
22.	eP	A	19 34 57	<u>Mariana Islands</u> 18.80 N 146.04 E
	ePP	A	39 04	H = 19 21 21.5 h = 103 km MB = 5.1
				D = 100.09 Az = 332.5 (NEIS)
				PV A traces
				PPV A 1.5s 25.1nm M = 5.6
22.	eP	A	23 33 09.5	<u>Greece-Bulgaria Border Region</u>
	e	A	33 13	41.26 N 23.03 E
	e	A	23 33	H = 23 30 15.0 h = normal MB=5.1 MS=4.4
	LmH	B	37.4	D = 12.28 Az = 323.7 (NEIS)
	LmV	B	38.4	P1V A 1.4s 23.3nm M = 5.1
				P2V A 1.8 57.5nm 5.4
				LmH B 9.7 2.4 μm 4.5
				LmV B 12 3.0 μm
23.	eP	A	05 26 45	<u>Fox Islands, Aleutian Is.</u>
				52.54 N 169.03 W
				H = 05 14 53.8 h = 42 km MB = 5.0 MS=4.3
				D = 77.19 Az = 359.6 (NEIS)
				PV A 1.0s 19.7nm M = 5.1

Day	Phase		h m s	Remarks
23.	ePKHKP	A	06 31 48	Fiji Islands Region 20.67 S 178.22 W H = 06 12 53.9 h = 490 km MB = 4.4 D = 149.10 Az = 347.8 (NEIS) PKHKPV A 1.7s 18.2nm
23.	ePKIKP	A	06 39 48.5	Solomon Islands 7.01 S 155.84 E H = 06 20 50.3 h = 70 km MB = 5.5 D = 127.31 Az = 332.1 (NEIS) PKIKPV A 1.2s 44.7nm
23.	eP	A	09 48 01.5	Honshu, Japan 39.70 N 141.35 E H = 09 36 00.7 h = 87 km MB = 4.4 D = 79.83 Az = 330.2 (NEIS) traces
23.	ePKIKP	A	14 47 57.5	New Guinea 4.17 S 142.55 E H = 14 29 21.8 h = 107 km MB = 5.4 D = 118.14 Az = 326.9 (NEIS)
24.	eP	A	15 24 58	Republic of South Africa 26.27 S 27.26 E H = 15 12 57.8 h = 16 km MB = 5.0 D = 77.81 Az = 349.9 (NEIS) traces
24.	eP	A	19 13 28	South of Honshu, Japan 32.81 N 136.99 E
	epP	A	14 57	H = 19 01 40.0 h = 393 km MB = 5.3
	ePP	A	16 43	D = 83.95 Az = 328.5 (NEIS)
	eS	B	23 17	h = 393 km
				PV A 1.4s 18.6nm M = 4.6
24.	ePKiPK	A	20 53 35	South Sandwich Islands Region
	ePP	AB	53 39	55.83 S 27.45 W
	eSKKS	B	21 00 36	H = 20 34 35.4 h = 80 km MB=6.0 (NEIS)
	ePS	B	03 08	D = 110.9
	ePKKP	A	04 06	PKiKPV A 1.9s 53.0nm
	LmH	B	33.6	PPV A 1.8 64.2nm M = 5.8

Day	Phase		h m s	Remarks
cont. 24.	LmV	B	21 33.6	PKKPV A 1.2s 12.2nm LmH B 23.5 3.0/um LmV B 22 2.5/um
24.	ePKIKP LmH LmV	A B B	21 53 52 22 44.9 47.3	Near North Coast of New Guinea 2.28 S 141.07 E H = 21 35 09.8 h = normal MB=5.7 MS=5.6 D = 115.77 Az = 326.9 (NEIS) LmH B 21s 2.0/um M = 5.7 LmV B 18 1.7/um 5.7
25.	eP	A	03 57 26	Hokkaido, Japan Region 44.50 N 144.55 E H = 03 45 54.1 h = 189 km MB = 4.7 D = 76.80 Az = 331.4 (NEIS)
25.	eP	A	04 04 46.5	Eastern Kazakh SSR 49.89 N 78.12 E H = 03 56 57.6 h = 0 km MB = 4.7 D = 41.21 Az = 297.6 (NEIS) Underground explosion (UPP) PV A 0.7s 15.3nm M = 4.8
25.	eP	A	05 13 51	Near Coast of Oaxaca, Mexico 15.48 N 95.38 W H = 05 00 58.9 h = 25 km MB=5.3 MS=4.9 D = 88.56 Az = 37.5 (NEIS)
25.	ePKIKP e(PKHKP) ePP LmH LmV	A A A B B	05 25 12 25 21 29 27 06 31.4 31.4	South Pacific Cordillera 54.64 S 131.62 W H = 05 05 19.0 h = normal MB=6.1 MS=5.7 D = 157.51 Az = 85.0 (NEIS) PKIKPV A 1.8s 40.5nm PKHKPV A 1.8 74.3nm LmH B 20 0.7/um M = 5.4 LmV B 20 1.3/um 5.8
25.	eP eS	A B	08 57 38 09 08 35	Near Coast of Oaxaca, Mexico 15.44 N 95.47 W

Day	Phase		h m s	Remarks
cont. 25.	LmH	B	09 38.9	H = 08 44 45.3 h = 30 km MB=5.6 MS=5.0
	LmV	B	38.9	D = 88.64 Az = 37.5 (NEIS)
				PV A 1.7s 39.4nm M = 5.5
				LmH B 18 0.97 /um 5.3
				LmV B 18 1.15 /um 5.4
25.	ePKP	A	16 49 40	Tonga Islands 16.05 S 173.27 W
				H = 16 30 04.4 h = normal MB = 4.6
				D = 145.26 Az = 354.5 (NEIS)
				traces
25.	eP1	AB	17 36 02	South Indian Ocean 26.09 S 84.28 E
	eP2	A	36 07	H = 17 22 19.3 h = normal MB=6.2 MS=6.6
	ePP	B	40 04	D = 99.60 Az = 321.9 (NEIS)
	eSKS	B	46 36	P1V A traces
	eS	B	47 40	P2V A 2.6s 138.7nm M = 6.4
	ePS	B	49 00	LmH B 20 10.4 /um 6.3
	LmH	B	18 20.6	LmV B 18.5 8.7 /um 6.3
	LmV	B	28.2	
25.	eP	A	17 52 26	Off East Coast of Honshu, Japan
				34.43 N 141.80 E
				H = 17 39 50.0 h = 9 km MB = 5.0
				D = 84.58 Az = 330.8 (NEIS)
25.	eP1	AB	22 28 25	Iceland 64.60 N 17.65 W
	eiP2	AB	28 30	H = 22 23 46.2 h = normal MB=5.1 MS=5.2
	eS1	B	32 20	D = 20.73 Az = 118.4 (NEIS)
	eS2	B	32 30	P1V A 1.5s 201.0nm M = 5.3
	LmH	B	38.5	P2V A 1.8 307.4nm 5.4
	LmV	B	38.9	S1H B 5.5 1.4 /um 5.3
				S2H B 8 2.6 /um 5.4
				LmH B 17 3.2 /um 4.8
				LmV B 14 3.5 /um 5.0
26.	eP	A	01 27 29	Svalbard Region 76.13 N 9.3 E
				H = 01 21 58.1 h = 0 km MB = 4.3
				D = 25.60 Az = 177 (ISC)
				PV A 1.4s 18.6nm M = 4.5

Day	Phase		h m s	Remarks
26.	ePKIKP	A	14 02 42.5	Southern Pacific Ocean 36.56 S 98.20 W
	eSS	C	22 30	H = 13 43 35.3 h = normal MB=5.4 MS=5.6
	eSSS	C	26 50	D = 129.13 Az = 50.5 (NEIS)
	LmH	B	59.0	LmH B 20s 0.6/um M = 5.3
	LmV	B	15 00.5	LmV B 20 0.6/um 5.3
26.	eP	A	14 29 31	Turkey 36.60 N 34.44 E
				H = 14 24 37.7 h = normal MB = 4.0
				D = 21.57 Az = 317.8 (NEIS)
26.	eP	A	18 53 28	North Atlantic Ridge 10.68 N 43.95 W
	LmH	B	19 17.0	H = 18 43 16.5 h = normal MB = 4.7
	LmV	B	17.0	D = 60.28 Az = 37.2 (NEIS)
				LmH B 18s 0.6/um M = 4.7
				LmV B 18 0.6/um 4.9
26.	eP	A	19 04 26	North Atlantic Ridge 10.96 N 43.33 W
				H = 18 54 17.9 h = normal MB=4.7 MS=4.8
				D = 59.68 Az = 37.1 (NEIS)
26.	ePKIKP	A	23 52 14.5	South of Fiji Islands 23.86 S 179.23 E
	ePKHKP	A	52 21	H = 23 33 28.7 h = 551 km MB = 5.4
	ePKP2	A	52 33	D = 151.58 Az = 343.3 (NEIS)
				PKIKPV A 1.6s 44.0nm
				PKHKPV A 1.2 89.4nm
				PKP2V A 1.4 51.1nm
27.	eP1	AB	02 01 39.5	South of Honshu, Japan 33.79 N 139.20 E
	eP2	AB	01 46	H = 01 49 08.1 h = 16 km MB=5.7 MS=5.9
	ePP1	A	04 54	D = 84.06 Az = 329.5 (NEIS)
	ePP2	AB	05 02	P1V A 1.8s 87.8nm M = 5.7
	eS	B	12 02	P2V A 1.6 236.3nm 6.2
	eSS	B	16 22	P2V B 4.5 1.2/um 6.4
	LmH	B	40.2	SH B 12 2.2/um 6.1
	LmV	B	46.9	LmH B 16.5 24.2/um 6.7
				LmV B 15 13.7/um 6.5

Day	Phase		h m s	Remarks
27.	ePKP	A	02 08 50	Fiji Islands Region 18.12 S 177.85 W
				H = 01 50 11.2 h = 591 km MB = 4.8
				D = 146.68 Az = 349.0 (NEIS)
27.	eP	AB	05 01 34	Shikoku, Japan 32.26 N 132.17 E
	epP	AB	01 45	H = 04 49 15.5 h = 39 km MB = 5.0
	LmV	B	42.4	D = 82.22 Az = 326.4 (NEIS)
	LmH	B	42.5	h = 38 km
				PV A 3.0s 131.6nm M = 5.4
				pPV A 2.7 76.7nm
				LmH B 15 3.3/um 5.8
				LmV B 15 3.2/um 5.8
27.	+iPKIKP	AB	08 05 03.5	New Britain Region 4.72 S 152.51 E
	epPKIKP	A	05 22	H = 07 46 11.9 h = 70 km MB = 6.1
	e(PKKS)	A	18 52	D = 123.74 Az = 331.1 (NEIS)
	LmH	B	59.7	h = 65 km
	LmV	B	59.7	PKIKPV A 1.7s 78.8nm
				LmH B 20 1.8/um
				LmV B 22 1.9/um
27.	ePKP2	A	17 08 32.5	South of Kermadec Islands
				33.28 S 178.45 W
				H = 16 47 51.20 h = normal
				MB = 4.9 MS = 5.3 (NEIS)
				D = 161.2
				PKP2V A 1.2s 12.2nm
27.	eP	A	18 56 36	Central Mid-Atlantic Ridge
				1.52 N 30.84 W
				H = 18 46 25.7 h = normal MB=5.4 MS=4.9
				D = 60.67 Az = 29.5 (NEIS)
				PV A 1.6s 22.0nm M = 5.0
27.	eP	A	19 04 24	Andaman Islands Region 10.50 N 92.79 E
				H = 18 52 36.2 h = normal MB = 5.3
				D = 76.36 Az = 319.7 (NEIS)
				PV A 1.2s 34.6nm M = 5.3

Day	Phase		h m s	Remarks
27.	eP	A	23 01 18	Central Mid-Atlantic Ridge
				1.69 N 30.84 W
				H = 22 51 09.2 h = normal MB=5.3 MS=5.0
				D = 60.52 Az = 29.6 (NEIS)
				PV A 1.4s 27.9nm M = 5.2
28.	ePKP2	A	02 31 22	South of Kermadec Islands
				33.43 S 178.91 W
				H = 02 10 46.09 h = 61 km MB = 5.0 (NEIS)
				D = 161.2
28.	ePKP2	A	02 53 30	South of Kermadec Islands
				33.28 S 178.52 W
				H = 02 32 50.26 h = 38 km MB = 5.0 (NEIS)
				D = 161.1
				PKP2V A 1.6s 16.5nm
28.	ePn	A	05 30 23.5	Hesse, Fed. Rep. of Germany
	ePg	A	30 29	51.58 N 7.79 E
	eSg	A	31 03	H = 05 29 39.5 h = 1 km
				D = 2.59 Az = 109.7 (NEIS)
				PnV A 0.7s 23.0nm
28.	eP1	A	11 13 09	Algeria 36.59 N 5.27 E
	eP2	AB	13 15	H = 11 09 40.3 h = normal MB=5.0 MS=4.8
	eSS	B	16 10	D = 14.78 Az = 16.0 (NEIS)
	LmH	B	18.1	P1V A 1.5s 17.6nm M = 4.3
	LmV	B	21.8	P2V A 1.4 74.5nm 4.9
				P2H A 1.5 64.4nm 4.9
				LmH B 17.5 1.6 µm 4.2
				LmV B 9 1.2 µm
29.	eP	AB	01 10 30	Algeria 36.67 N 5.19 E
	eSS	B	13 34	H = 01 06 58.6 h = normal MB=4.7 MS=4.5
	LmH	B	17.3	D = 14.72 Az = 16.3 (NEIS)
	LmV	B	19.2	PV A 1.4s 32.6nm M = 4.6
				PH A 1.5 30.7nm 4.6
				LmH B 15 1.6 µm 4.3
				LmV B 9 1.6 µm

Day	Phase		h m s	Remarks
29.	e	A	01 39 32	
29.	eP	A	21 50 11	Off East Coast of Kamchatka 52.00 N 161.23 E H = 21 38 32.9 h = 29 km MB = 4.6 D = 74.54 Az = 340.5 (NEIS) traces
29.	e e LmH	A A B	22 35 02 35 12 39.2	Southern Italy 39.65 N 18.66 E H = 22 32 00.7 h = normal MB = 4.0 D = 12.06 Az = 338.1 (NEIS) LmH B 14s 0.38/um M = 3.53
30.	ePKHKP ePKIKP LmH LmV	A A B B	08 53 08 53 10.5 09 56.2 56.2	New Hebrides Islands 17.99 S 168.29 E H = 08 33 46.5 h = 61 km MB = 5.7 D = 142.48 Az = 335.5 (NEIS) PKIKPV A 1.6s 49.5nm LmH B 22 1.4/um LmV B 22 1.1/um
30.	eP	A	17 21 39	Malagasy Republic 13.55 S 48.78 E H = 17 10 15.0 h = normal MB=4.5 MS=5.4 D = 71.70 Az = 336.1 (NEIS)
30.	ePKIKP	A	18 14 45	Solomon Islands 7.12 S 155.75 E H = 17 55 44.4 h = 53 km MB = 5.3 D = 127.37 Az = 332.0 (NEIS) PKIKPV A 1.3s 21.8nm
30.	eP	A	18 36 29.5	Near East Coast of Kamchatka 53.92 N 160.80 E H = 18 25 03.9 h = normal MB = 4.8 D = 72.64 Az = 340.0 (NEIS) PV A 1.0s 15.7nm M = 5.0
30.	ePg eSn eSg	A A A	19 07 33 08 08 08 52.5	Northern Italy 44.15 N 10.68 E H = 19 05 23.8 h = normal MB = 4.3 D = 6.54 Az = 5.3 (NEIS)

Day	Phase		h m s	Remarks
1.	ePn	A	01 27 37.5	France 49.42 N 6.03 E
	ePg	A	27 51	H = 01 26 38.0 h = 0 km MB = 4.3
	ei	A	28 27.0	D = 3.80 Az = 69.0 (NEIS)
	eiSg	A	28 49.0	PnV A 1.2s 16.3nm M = 4.7
1.	eP	A	05 18 09	Greenland Sea 79.86 N 1.15 W
	e	A	18 17	H = 05 12 04.8 h = normal MB = 4.2
				D = 29.66 Az = 163.5 (NEIS)
				PV A 1.0s 13.8nm
1.	eP	A	05 31 19.5	East of Lake Baikal 56.39 N 113.66 E
	e	A	31 27	H = 05 21 46.0 h = normal MB = 4.6
				D = 55.46 Az = 310.9 (NEIS)
1.	ePKIKP	A	06 42 35.5	Fiji Islands Region 19 S 176 W
	ePKHKP	A	42 38	H = 06 22 58 MB = 4.8 (NORSAR)
				D = 147.8
1.	eP	A	07 33 58	Greenland Sea 75.72 N 6.82 E
	e	A	34 04.5	H = 07 28 31.2 h = normal MB=4.6 MS=5.0
				D = 25.24 Az = 172.8 (NEIS)
1.	eP	AB	17 05 32	Salta Province, Argentina
	ePP	B	09 35	22.14 S 64.74 W
	eSKS	B	16 11	H = 16 51 51.5 h = 13 km MB=5.5 MS=5.9
	eS	B	17 00	D = 98.60 Az = 38.7 (NEIS)
	ePS	B	18 25	PV A traces
	LmV	B	48.3	PV B 8s 0.5/um M = 6.1
	LmH	B	52.7	PPV B 8 0.5/um 6.0
				SH B 13 1.1/um 6.2
				LmH B 17.8 4.6/um 6.0
				LmV B 19.5 6.9/um 6.2
1.	eP	AB	23 22 57	South Atlantic Ridge 22.64 S 10.67 W
	eS	B	32 38	H = 23 11 14.5 h = normal MB=5.6 MS=5.6
	LmH	B	56.2	D = 75.58 Az = 14.4 (NEIS)
	LmV	B	56.3	PV A 1.3s 48.0nm M = 5.3
				LmH B 17.5 2.1/um 5.5
				LmV B 17 2.6/um 5.6

Day	Phase		h m s	Remarks
2.	e(PKHKP)	A	06 50 35	South of Fiji Islands 26 S 180 W H = 06 30 34 h = 3.6 (NORSAR) D = 153.8
2.	ePKHKP	A	07 35 33	West of Macquarie Island 54.14 S 140.21 E H = 07 15 46.1 h = normal MB = 5.4 MS=5.3 D = 149.00 Az = 284.9 (NEIS) PKHKPV A 1.6s 33.0nm
2.	iPn eiSg	A A	08 02 13 02 52.5	Lower Silesia, Poland (CLL)
2.	eP LmH LmV	A B B	16 49 09 17 07.9 08.4	Alma-Ata Region 42.17 N 75.56 E H = 16 41 05.8 h = normal MB = 5.0 D = 43.65 Az = 304.0 (NEIS) LmH B 14s 1.0/um M = 4.9 LmV B 13 0.7/um 4.9
2.	LmH LmV	B B	20 35.8 35.8	Near East Coast of Honshu 37.75 N 141.78 E H = 19 19 43.8 h = 86 km MB = 4.3 (ISC) D = 81.5 LmV B 18s 0.4/um
2.	eP	A	20 37 11	Off East Coast of Honshu, Japan 40.57 N 143.62 E H = 20 25 02.8 h = 28 km MB = 4.7 D = 79.92 Az = 331.3 (NEIS)
2.	ePKIKP ePKHKP ePKP2 LmV LmH	AB AB A B B	23 46 20.5 46 35 46 54 24 49.1 49.2	Kermadec Islands Region 29.08 S 175.95 W H = 23 26 26.6 h = normal MB=6.8 MS=7.2 D = 157.72 Az = 347.2 (NEIS) PKIKPV A 3.0s 2580.0nm PKIKPV B 9 9.3nm LmH B 21.5 63.5/um M = 7.3 LmV B 26 106.0/um 7.5

Day	Phase		h m s	Remarks
3.	eP	A	05 13 18	Off Coast of Northern California
				40.42 N 125.14 W
				H = 05 00 58.6 h = 12 km MB=5.4 MS=5.2
				D = 81.72 Az = 26.2 (NEIS)
				PV A 1.3s 21.8nm M = 5.1
3.	ePKIKP	AB	23 45 02.5	Kermadek Islands Region
	ePKHKP	A	45 15.5	29.12 S 176.11 W
	ePKP2	A	45 35.5	H = 23 25 09.3 h = normal MB=6.2 MS=6.6
	ePP	AB	49 12.5	D = 157.73 Az = 347.0 (NEIS)
	e(PPP)	B	52 50	PKIKPV A 3.0s 1110.0nm
	ePPS	B	24 02 40	PKIKPV B 12 4.4/um
	LmV	B	25 08.5	LmH B 17.8 9.0/um M = 5.5
	LmH	B	09.5	LmV B 17.5 7.0/um 5.5
4.	ePKP2	A	13 54 45	Kermadec Islands Region
				29.13 S 146.14 W
				H = 13 34 18.85 h = normal
				MB = 5.1 MS = 5.0 (NEIS)
				D = 157.8
				PKP2V A 1.2s 16.3nm
4.	+iP1	AB	19 39 56	Mongolia 45.14 N 94.03 E
	iP2	A	40 02	H = 19 30 42.1 h = normal MB=6.1 MS=6.7
	ePP	B	42 00	D = 52.83 Az = 307.6 (NEIS)
	eS	B	47 26	P1V A 1.2s 228.0nm M = 6.0
	eSS	B	51 08	P2V A 1.8 764.0nm 6.3
	LmH	B	20 03.3	LmH B 13 202.0/um 7.4
	LmV	B	05.4	LmV B 12.5 169.0/um 7.4
4.	eP	A	20 11 04	Northern Sinkiang Province
				44.6 N 93.90 E
				H = 20 01 41.4 h = 58 km MB = 4.5
				D = 53.06 Az = 308
4.	eP	A	21 35 24	Hindu Kush Region 36.39 N 70.85 E
				H = 21 27 33.24 h = 202 km MB = 4.6 (NEIS)
				D = 43.9

Day	Phase		h m s	Remarks
5.	LmV	B	06 08.6	<u>Off Coast of Southern Chile</u>
	LmH	B	10.8	44.78 S 80.01 W
				H = 04 58 55.44 h = normal
				MB = 5.2 MS = 5.1 (NEIS)
				D = 123.6
				LmH B 19s 0.7/um M = 5.3
				LmV B 21 1.0/um 5.4
5.	ePKP2	A	18 24 41	<u>Kermadek Islands Region</u>
	eX	A	24 48.5	29.23 S 176.25 W
	e	A	25 03.5	H = 18 04 14.92 h = normal
	ePP	A	28 44	MB = 5.4 MS = 5.3 (NEIS)
	LmV	B	19 42.6	D = 157.7
	LmH	B	42.8	XV A 1.8s 54.1nm
				PPV A 2.0 59.8nm M = 5.4
				LmV B 17 0.35/um 5.3
5.	eP	A	20 21 50	<u>Caribbean Sea</u> 14.75 N 81.73 W
				H = 20 09 34.6 h = 23 km MB = 5.2
				D = 80.84 Az = 40.1 (NEIS)
6.	LmH	B	13 32.7	<u>Easter Island Cordillera</u>
	LmV	B	36.6	34.97 S 107.80 W
				H = 12 20 59.4 h = 33 km MB = 5.2 (ISC)
				D = 134.2
				LmH B 19s 0.4/um M = 5.1
				LmV B 18 0.5/um 5.3
6.	LmH	B	15 52.7	<u>Scotia Sea</u> 60.76 S 38.09 W
	LmV	B	16 06.3	H = 14 51 23.61 h = normal
				MB = 5.6 MS = 5.5 (NEIS)
				D = 112.6
				LmH B 24s 1.1/um M = 5.4
				LmV B 17 0.9/um 5.4
6.	eP	A	20 25 10.5	<u>Shikoku, Japan</u> 33.61 N 132.34 E
				H = 20 13 00.7 h = 65 km MB = 4.8
				D = 81.18 Az = 326.4 (NEIS)

Day	Phase		h m s	Remarks
6.	eP	A	20 30 21	Mongolia 45.12 N 94.09 E H = 20 21 05.5 h = normal MB = 4.4 D = 52.87 Az = 307.7 (NEIS)
6.	eP	A	21 38 00	South Atlantic Ridge 22.57 S 10.70 W H = 21 26 18.3 h = normal MB = 5.0 D = 75.51 Az = 14.5 (NEIS)
6.	ePKP2	A	23 39 51	South of Fiji Islands 26.87 S 178.46 W H = 23 20 10.6 h = 327 km MB = 4.8 D = 155.04 Az = 344.7 (NEIS) PKP2V A 1.3s 10.9nm
7.	eP	A	08 31 52.5	Northern Sumatra 5.18 N 96.30 E H = 08 19 46.1 h = 169 km MB = 4.9 D = 82.65 Az = 320.3 (NEIS) PV A 1.5s 25.1nm M = 4.8
7.	eP e	A A	13 05 11.5 05 29.5	Kurile Islands 46.63 N 152.72 E H = 12 53 17.3 h = 33 km MB = 5.2 D = 77.45 Az = 335.8 (NEIS) PV A 1.2s 36.6nm M = 5.6
7.	ePKHKP	A	15 32 37	Tonga 21.7 S 174.7 W H = 15 12 49 h = 33 km MB = 4.7 D = 150.69 Az = 352 (ISC) PKHKPV A 1.8s 20.3nm
7.	eP	A	21 06 04	Northern India 30.64 N 78.69 E H = 20 56 49.7 h = normal MB = 4.9 D = 52.79 Az = 312.6 (NEIS)
8.	ePn eSg	A A	00 18 50.5 20 05	Yugoslavia 46.2 N 15.5 E H = 00 17 23 (BCIS) D = 5.14
8.	+iP1 eP2 ePP	AB A B	05 57 57.8 58 09 06 01 10	Near East Coast of Honshu, Japan 36.41 N 141.10 E H = 05 45 37.0 h = 35 km MB=6.0 MS=6.0

Day	Phase		h m s	Remarks
cont. 8.	eS	B	06 08 10	D = 82.58 Az = 330.3 (NEIS)
	LmH	B	37.5	P1V A 1.6s 269.2nm M = 6.0
	LmV	B	39.9	P1V B 8 3.2/um 6.4
				P2V A 2.0 376.1nm 6.1
				PPV B 7.0 1.4/um 6.4
8.	eP	A	06 05 41	Ural Mountains Region 53.68 N 55.1 E
	e	A	06 04	H = 05 59 59.8
				D = 26.53 Az = 281 (ISC)
8.	eP	A	12 21 20	Near East Coast of Kamchatka
				52.57 N 158.83 E
				H = 12 09 52.4 h = 60 km MB = 4.8
				D = 73.49 Az = 338.9 (NEIS)
				PV A 1.2s 16.2nm M = 4.8
9.	eP	AB	02 36 30.5	Dodecanese Islands 36.65 N 28.45 E
	e	A	36 39.5	H = 02 32 17.6 h = 69 km MB = 5.0
	e	A	36 43	D = 18.49 Az = 324.4 (NEIS)
	LmH	B	44.8	PV A 1.4s 237.0nm M = 5.2
	LmV	B	44.9	PV B 4 0.8/um 5.3
				LmH B 14 1.0/um
				LmV B 14 1.1/um
9.	e	A	03 05 42	
9.	eP	A	16 31 40	Szechwan Province, China
				28.23 N 103.89 E
				H = 16 20 29.5 h = 33 km MB = 5.2
				D = 70.09 Az = 317.4 (NEIS)
				PV A 1.2s 12.2nm M = 4.8
9.	ePKP2	A	17 05 17	Kermadec Islands Region
				29.33 S 176.21 W
				H = 16 44 50.68 h = normal MB = 4.7 (NEIS)
				D = 157.9

Day	Phase		h m s	Remarks
9.	ePcP	A	17 49 48	Caribbean Sea 14.61 N 81.81 W H = 17 37 21.9 h = normal MB=5.0 MS=4.9 D = 81.00 Az = 40.0 (NEIS)
10.	ePKP	A	00 23 55	Tonga Islands 16.47 S 173.81 W H = 00 04 17.5 h = 31 km MB = 4.5 D = 145.63 Az = 353.9 (NEIS) PKPV A 0.9s 15.6nm
10.	+iP	A	03 04 46.4	Eastern Kazakh SSR 49.79 N 78.14 E
	ePn	A	06 19	H = 02 56 57.5 h = 0 km MB = 5.3 D = 41.27 Az = 297.7 (NEIS) Underground explosion M = 5.2 (UPP) PV A 0.7s 61.3nm M = 5.4
10.	ePcP	A	04 42 32.5	Off W. Coast of Northern Sumatra 0.67 N 96.70 E H = 04 29 41.5 h = normal MB = 4.9 D = 86.36 Az = 320.5 (NEIS)
10.	eP	A	12 19 25.5	Dodecanese Islands 36.14 N 28.29 E H = 12 15 05.9 h = 45 km MB = 3.9 D = 18.83 Az = 325.5 (NEIS) PV A 1.8s 27.0nm M = 4.2
10.	+eP	AB	16 12 17.5	Southern Nevada 37.07 N 116.03 W
	ePP	A	15 23	H = 16 00 00.1 h = 0 km MB = 5.7 D = 81.27 Az = 30.7 (NEIS) Underground explosion (AEC) PV A 1.5s 57.8nm M = 5.4
11.	LmH	B	06 38.3	Mariana Islands 13.10 N 145.38 E
	LmV	B	44.7	H = 05 34 22.10 h = 53 km MB = 5.3 (NEIS) D = 104.9 LmH B 16s 0.4/um LmV B 17 0.5/um

Day	Phase		h m s	Remarks
11.	eP	A	16 06 00.5	Hokkaido, Japan Region 41.95 N 142.30 E
				H = 15 54 07.2 h = 73 km MB = 4.7
				D = 78.23 Az = 330.5 (NEIS)
				PV A traces
11.	+eP	A	18 01 14	Jan Mayen Island Region 71.57 N 4.13 W
	eS	B	05 18	H = 17 56 18.5 h = normal MB = 5.0
	LmH	B	10.2	D = 22.18 Az = 152.8 (NEIS)
	LmV	B	11.2	PV A 2.0s 162.0nm M = 5.1
				SH B 9 0.8 µm 5.0
				LmH B 17.5 0.8 µm 4.2
				LmV B 14 0.6 µm 4.3
11.	ePKHKP	A	19 30 53.5	South of Fiji Islands 22.88 S 176.48 W
	ePKP2	A	31 04	H = 19 11 24.9 h = 222 km MB = 4.6
				D = 151.58 Az = 349.1 (NEIS)
				traces
13.	+eiP1	AB	01 30 53	Panama-Colombia Border Region
	eiP2	A	31 14.5	7.75 N 77.69 W
	eS	B	41 18	H = 01 18 22.8 h = 12 km MB = 7.3
	ePS	B	42 16	D = 83.63 Az = 39.8 (NEIS)
	eSS	B	47 15	P1V A 1.4s 288.0nm M = 6.3
	eLQ	B	53 00	P1V B 8 13.6 µm 7.2
	P'P'	A	57 25	P2V A 2.2 1319.5nm 6.7
	e	A	57 29.5	SH B 13 2.9 µm 6.2
	LmH	B	02 04.7	LmH B 19.8 108.0 µm 7.2
	LmV	B	04.8	LmV B 19.5 142.0 µm 7.4
13.	eP	A	01 43 25	Panama-Colombia Border Region
				7.54 N 77.82 W
				H = 01 30 56.8 h = normal MB = 5.4
				D = 83.87 Az = 39.8 (NEIS)
				PV A 1.3s 21.8nm M = 5.1

Day	Phase		h m s	Remarks
13.	e(P)	A	01 52 38	Near West Coast of Colombia 6.94 N 77.86 W H = 01 39 59.3 h = 21 km MB = 5.2 D = 84.35 Az = 39.8 (NEIS) (P)V A 1.3s 17.5nm M = 5.1
13.	eP	A	02 09 09	Panama-Colombia Border Region 7.61 N 77.77 W H = 01 56 39.9 h = normal MB = 4.8 D = 83.79 Az = 39.8 (NEIS) PV A traces
13.	eP	A	02 19 43	Panama-Colombia Border Region 7.58 N 77.67 W H = 02 07 13.1 h = 24 km MB = 5.3 D = 83.75 Az = 39.8 (NEIS) PV A 1.6s 33.0nm M = 5.3
13.	eP	A	02 25 08	Panama-Colombia Border Region 7.63 N 77.74 W H = 02 12 39.6 h = normal MB = 5.0 D = 83.75 Az = 39.8 (NEIS) PV A traces
13.	eP	A	02 27 22.5	Panama-Colombia Border Region 7.55 N 77.65 W H = 02 14 54.0 h = 23 km MB = 5.2 D = 83.76 Az = 39.8 (NEIS)
13.	eP	A	02 32 50	Panama-Colombia Border Region 7.41 N 77.66 W H = 02 20 22.8 h = 45 km MB = 5.5 (NEIS) D = 83.9 PV A 1.3s 35.0nm M = 5.2
13.	eP	A	03 11 14	Panama-Colombia Border Region 7.40 N 77.79 W H = 02 58 45.6 h = normal MB = 4.9 D = 83.96 Az = 39.8 (NEIS)

Day	Phase		h m s	Remarks
13.	eP	A	04 03 34.5	**Panama-Colombia Border Region**
	e	A	03 40	7.75 N 77.70 W
				H = 03 51 04.0 h = 23 km MB = 5.0
				D = 83.64 Az = 39.8 (NEIS)
				PV A 1.4s 23.2nm M = 5.2
13.	eP	A	05 07 26	**Panama** 7.70 N 78.26 W
				H = 04 54 58.3 h = normal MB = 4.4
				D = 84.03 Az = 39.8 (NEIS)
				PV A traces
13.	eP	A	10 33 57	**Panamo-Colombia Border Region**
				7.73 N 77.61 W
				H = 10 21 24.3 h = 12 km MB = 4.9
				D = 83.60 Az = 39.8 (NEIS)
13.	eP	A	10 50 27.5	**Northeast of Taiwan** 26.55 N 125.68 E
				H = 10 38 06.8 h = 97 km MB = 5.1
				D = 83.67 Az = 324.2 (NEIS)
				PV A 1.5s 25.1nm M = 4.9
13.	e	A	12 55 51	**Southern Alaska** 61.49 N 145.01 W
				H = 12 44 50.7 h = 55 km MB = 4.7
				D = 66.66 Az = 16.0 (NEIS)
13.	eP	A	13 00 10	**Panama-Colombia Border Region**
				7.68 N 77.52 W
				H = 12 47 39.6 h = 29 km MB=4.9 MS=4.2
				D = 83.58 Az = 39.8 (NEIS)
				PV A 1.3s 13.1nm M = 4.9
13.	eP	A	13 13 21	**Panama-Colombia Border Region**
	e	A	13 35	7.47 N 77.62 W
				H = 13 00 53.3 h = normal MB=4.9 MS=4.6
				D = 83.80 Az = 39.8 (NEIS)
13.	eP	A	16 01 03.5	**Algeria** 35.97 N 4.79 E
	eiX	A	01 08	H = 15 57 25.2 h = 37 km MB = 4.8

Day	Phase		h m s	Remarks
cont. 13.	LmH	B	16 06.4	D = 15.48 Az = 16.5 (NEIS)
	LmV	B	07.8	XV A 1.2s 61.0nm
				LmH B 17.5 2.2/um M = 4.4
				LmV B 14 1.2/um 4.4
13.	eP	A	16 28 25	<u>Panama-Colombia Border Region</u>
				7.78 N 77.64 W
				H = 16 15 56.0 h = 18 km MB = 4.9
				D = 83.57 Az = 39.8 (NEIS)
13.	eP	A	16 47 49	<u>Panama-Colombia Border Region</u>
				7.35 N 77.93 W
				H = 16 35 19.8 h = 37 km MB = 4.7
				D = 84.09 Az = 39.8 (NEIS)
				PV A 1.2s 10.2nm M = 4.9
13.	eP	A	18 11 13	<u>Panama-Colombia Border Region</u>
	eS	B	21 36	7.70 N 77.72 W
	LmH	B	49.5	H = 17 58 41.4 h = 5 km MB=5.4 MS=5.7
	LmV	B	49.5	D = 83.68 Az = 39.8 (NEIS)
				PV A 2.6s 139.0nm M = 5.7
				PV B 8 0.7/um 5.9
				SH B 15 0.9/um 5.7
				LmH B 18 1.2/um 5.3
				LmV B 18 2.3/um 5.6
13.	ePKP	A	18 17 59	<u>New Hebrides Islands</u> 20.79 S 169.30 E
	e	A	18 09.5	H = 17 58 25.3 h = 46 km MB = 5.1
				D = 145.41 Az = 334.8 (NEIS)
				PKPV A 1.2s 22.4nm
13.	ePKP	A	19 03 52	<u>Tonga Islands</u> 17.52 S 173.31 W
				H = 18 44 10.6 h = normal MB = 4.7
				D = 146.72 Az = 354.3 (NEIS)
				PKPV A 2.0s 34.2nm

Day	Phase		h m s	Remarks
13.	eP	A	19 28 14	<u>Panama-Colombia Border Region</u> 7.75 N 77.60 W H = 19 15 45.6 h = 18 km MB = 4.6 D = 83.57 Az = 39.8 (NEIS)
13.	eP1 eP2 LmH LmV	A B B	23 21 13 21 15 24 01.0 01.0	<u>Panama-Colombia Border Region</u> 7.15 N 77.73 W H = 23 08 41.9 h = 23 km MB=5.3 MS=4.3 D = 84.12 Az = 39.8 (NEIS) P1V A 1.5s 12.6nm M = 4.9 P2V A 1.3 26.2nm 5.3
14.	eP	A	02 01 11	<u>Panama-Colombia Border Region</u> 7.81 N 77.64 W H = 01 48 43.6 h = 26 km MB=5.2 MS=4.5 D = 83.55 Az = 39.8 (NEIS) PV A 1.4s 32.6nm M = 5.3
14.	eP eS LmH LmV	A B B B	02 26 19 36 44 03 05.6 11.5	<u>Panama-Colombia Border Region</u> 7.75 N 77.61 W H = 02 13 50.4 h = 15 km MB=5.9 MS=5.0 D = 83.59 Az = 39.8 (NEIS) PV A 1.7s 91.0nm M = 5.7 LmH B 17 0.5 μm 5.0 LmV B 18 0.6 μm 5.1
14.	eP	A	02 59 45	<u>Strait of Gibraltar</u> 35.59 N 3.74 W H = 02 55 26.8 h = 31 km MB = 4.4 D = 18.70 Az = 31.7 (NEIS)
14.	e	A	03 02 54	<u>Panama-Colombia Border Region</u> 7.74 N 77.40 W H = 02 50 15.1 h = 17 km MB = 4.8 D = 83.45 Az = 39.8 (NEIS)
14.	eSg	A	04 04 59	<u>Austria</u> 47.8 N 14.15 E H = 04 03 09 (VIE) D = 3.27

Day	Phase		h m s	Remarks
14.	eP	A	06 18 17.5	Greenland Sea 79.80 N 0.93 W
	e	A	18 25.5	H = 06 12 11.1 h = 17 km MB = 4.6
				D = 29.59 Az = 163.7 (NEIS)
				PV A 1.3s 13.1nm M = 4.6
14.	ePKP2	A	07 25 20	Kermadec Islands Region
				31.25 S 177.64 W
				H = 07 04 32.5 h = 15 km MB = 4.4 (NEIS)
				D = 159.4
				traces
14.	ePP	A	09 56 32	Flores Island Region 8.72 S 122.56 E
				H = 09 37 36.8 h = 123 km MB = 5.8 (NEIS)
				D = 109.9
				PPV A 1.6s 16.5nm M = 5.2
14.	eP	A	15 37 13.5	South of Honshu, Japan 33.28 N 139.79 E
				H = 15 24 53.4 h = 133 km MB = 4.6
				D = 84.75 Az = 329.8 (NEIS)
				PV A 1.0s 11.8nm M = 4.7
14.	eP	A	16 36 15	Northwest of Kurile Islands
				51.94 N 152.42 E
				H = 16 25 26.8 h = 378 km MB = 4.5
				D = 72.54 Az = 335.0 (NEIS)
				traces
14.	ePKIKP	A	19 08 16	Kermadec Islands Region
	ePKP2	A	08 46	28.27 S 178.15 W
	epPKP2	A	09 32	H = 18 48 42.9 h = 183 km MB = 5.3
				D = 156.45 Az = 344.3 (NEIS)
				h = 175 km
				PKP2V A 1.4s 27.9nm
				pPKP2V A 1.0 15.7nm
14.	eP	A	21 34 23	Greenland Sea 73.24 N 6.84 E
	eX	A	34 31	H = 21 29 21.8 h = normal MB = 4.7
				D = 22.78 Az = 172.1 (NEIS)
				PV A 1.5s 10.1nm M = 4.1
				XV A 1.4 20.9nm

Day	Phase		h m s	Remarks
15.	eP	A	09 17 58	Kamchatka 55.55 N 160.76 E H = 09 06 56.1 h = 161 km MB = 4.7 D = 71.10 Az = 339.8 (NEIS) PV A 1.2s 20.3nm M = 4.8
15.	eP	A	19 42 42.5	Panama-Colombia Border Region 7.87 N 77.18 W H = 19 30 14.8 h = 40 km MB = 4.5 D = 83.22 Az = 39.9 (NEIS)
15.	eP	A	23 23 57	Panama-Colombia Border Region 7.46 N 77.65 W H = 23 11 27.8 h = 28 km MB=5.3 MS=4.4 D = 83.83 Az = 39.8 (NEIS) PV A 1.3s 21.8nm M = 5.2
16.	ePKHKP	A	07 09 23	Kermadec Islands Region 29.10 S 176.32 W H = 06 49 21.1 h = 60 km MB = 4.7 D = 157.66 Az = 346.6 (NEIS) PKHKPV A 2.0s 29.9nm
16.	eP	A	24 05 49.5	Southern Sumatra 4.90 S 103.14 E H = 23 52 34.1 h = 58 km MB = 5.4 D = 94.72 Az = 320.3 (NEIS) traces
17.	eP	A	02 55 49	Near East Coast of Kamchatka 54.21 N 161.11 E H = 02 44 24.1 h = normal MB = 4.3 D = 72.43 Az = 340.2 (NEIS)
17.	ePKHKP	A	04 11 05	Tonga Islands 20.78 S 174.28 W H = 03 51 16.0 h = normal MB=4.7 MS=4.3 D = 149.84 Az = 352.5 (NEIS) PKHKPV A 1.6s 33.0nm

Day	Phase		h m s	Remarks
17.	eP	A	05 11 59	<u>Rumania</u> 45.75 N 26.53 E
	LmH	B	16.9	H = 05 09 23.0 h = 145 km MB = 5.1
	LmV	B	18.4	D = 11.09 Az = 301.6 (NEIS)
				PV A 1.6s 27.4nm M = 4.6
				LmH B 11 0.7/um
				LmV B 7 0.4/um
17.	ePKHKP	A	06 52 12	<u>Tonga Islands</u> 20.66 S 174.02 W
	LmV	C	08 25.0	H = 06 32 24.4 h = normal MB=4.8 MS=4.7
	LmH	C	28.0	D = 149.74 Az = 352.9 (NEIS)
				PKHKPV A 1.4s 27.9nm
				LmH C 15 0.35/um M = 5.2
				LmV C 16 0.35/um 5.2
17.	eP	A	10 54 38	<u>Andreanof Islands, Aleutian Is.</u>
	epP	A	54 50	51.66 N 173.51 W
	LmV	B	11 01.3	H = 10 42 42.3 h = 45 km MB = 5.0
	LmH	B	01.9	D = 77.98 Az = 356.7 (NEIS)
				h = 41 km
				PV A 1.2s 24.4nm M = 5.1
				LmH B 16 0.25/um 4.6
				LmV B 15 0.35/um 4.8
18.	eP	A	00 28 36	<u>Off East Coast of Kamchatka</u>
	LmH	B	01 03.2	52.42 N 160.29 E
	LmV	B	07.6	H = 00 17 02.0 h = normal MB=4.7 MS=4.3
				D = 73.95 Az = 339.9 (NEIS)
18.	eP	A	00 43 10	<u>Off East Coast of Kamchatka</u>
				52.37 N 160.14 E
				H = 00 31 36.4 h = normal MB = 4.4
				D = 73.96 Az = 339.8 (NEIS)
18.	LmV	C	03 56.5	<u>West Chile Rise</u> 41.5 S 87.1 W
	LmH	C	57.5	H = 02 46 10 h = 33 km MB = 4.6 (ISC)
				D = 112.8
				LmH C 25s 0.45/um M = 5.0
				LmV C 25 0.5/um 5.0

Day	Phase		h m s	Remarks
18.	ePKP	AB	11 24 17.5	Tonga Islands 15.22 S 173.59 W
	iX	A	24 20.5	H = 11 04 43.2 h = normal MB=5.9 MS=5.8
	eY	AB	24 31	D = 144.41 Az = 354.3 (NEIS)
	LmH	B	12 24.5	PKPV A 1.2s 32.5nm
	LmV	B	35.5	XV A 1.6 85.1nm
				YV A 1.6 82.4nm
				YV B 10 1.2/um
				LmH B 21 2.4/um M = 5.9
				LmV B 18.5 2.4/um 6.0
18.	e(Pn)	A	16 57 23	Northern Italy 44.8 N 7.9 E
	ePg	A	58 46	H = 16 55 38 h = 33 km
				D = 6.37 Az = 22 (ISC)
18.	eP	A	19 34 15.5	Guerrero, Mexico 17.06 N 98.35 W
	epP	A	34 29.5	H = 19 21 24.6 h = 48 km MB=5.6 MS=5.2
	ePP	A	37 47	D = 89.03 Az = 36.8 (NEIS)
				h = 49 km
				PV A 1.6s 38.5nm M = 5.5
18.	LmH	B	19 38.0	Off Coast of Southern Chile
	LmV	B	38.8	45.78 S 76.35 W
				H = 18 29 50.66 h = normal MB=5.1 MS=5.4
				D = 122.35 Az = 48.9 (NEIS)
				LmH B 20s 1.3/um M = 5.6
				LmV B 21 1.6/um 5.7
18.	ePKHKP	A	22 17 23	Fiji Islands Region 19.86 S 176.20 W
	epPKP	A	18 25	H = 21 58 09.2 h = 297 km MB = 4.6
				D = 148.66 Az = 350.4 (NEIS)
				traces
18.	e	A	22 25 41.5	Eastern Caucasus 42.55 N 45.34 E
				H = 22 20 11.7 h = normal MB = 4.5
				D = 24.38 Az = 301.1 (NEIS)
18.	e(Sg)	A	22 30 29	Upper Silesia (CLL)

Day	Phase		h m s	Remarks
18.	ePKIKP	A	23 30 56	Fiji Islands Region 20.66 S 178.43 W
	+iPKHKP	A	31 01.5	H = 23 12 18.4 h = 600 km MB = 4.6
	ePKP2	A	31 08	D = 149.05 Az = 347.5 (NEIS)
				PKHKPV A 1.4s 25.6nm
19.	LmH	B	03 06.5	Near Coast of Central Chile
	LmV	B	08.8	32.88 S 71.95 W
				H = 02 02 50.4 h = 36 km MB = 5.1 (ISC)
				D = 110.9
				LmV B 20s 0.35/um M = 5.0
19.	eP	A	07 45 00.5	Cyprus 35.73 N 31.56 E
				H = 07 40 23.3 h = 65 km MB = 4.2
				D = 20.72 Az = 322.1 (NEIS)
19.	ePKIKP	A	18 04 28.5	Solomon Islands 6.07 S 154.93 E
	ePP	AB	06 26	H = 17 45 43.9 h = 157 km MB = 5.7
				D = 126.06 Az = 331.9 (NEIS)
				PPV A 2.6s 130.0nm M = 5.5
19.	ePKIKP	A	18 52 58	Samoa Islands Region 16.28 S 171.92 W
				H = 18 33 21.7 h = normal MB=5.3 MS=4.8
				D = 145.60 Az = 356.0 (NEIS)
				PKIKPV A 1.5s 50.2nm
19.	eP	A	18 56 51	Hokkaido, Japan Region 42.20 N 142.63 E
				H = 18 44 53.4 h = 38 km MB = 4.8
				D = 78.14 Az = 330.6 (NEIS)
				traces
19.	eP	A	19 05 37.5	Northern Sumatra 3.48 N 98.27 E
				H = 18 53 01.8 h = normal MB = 4.8
				D = 85.21 Az = 320.4 (NEIS)
20.	eP	A	00 59 59.5	Andreanof Islands, Aleutian Is.
				51.59 N 173.55 W
				H = 00 48 04.0 h = 45 km MB=4.9 MS=4.2
				D = 78.04 Az = 356.6 (NEIS)

Day	Phase		h m s	Remarks
20.	eP	A	01 16 29.5	<u>Andreanof Islands, Aleutian Is.</u> 51.83 N 173.49 W H = 01 04 33.8 h = normal MB = 4.3 D = 77.81 Az = 356.7 (NEIS)
20.	ePKHKP	A	11 09 23	<u>Fiji Islands Region</u> 17.77 S 178.72 W H = 10 50 49.6 h = 624 km MB = 4.5 D = 146.17 Az = 348.2 (NEIS)
20.	eP	A	18 24 15.5	<u>Greece</u> 39.06 N 21.47 E H = 18 21 09.9 h = 43 km MB = 3.3 D = 13.51 Az = 332.2 (NEIS)
20.	e	A	19 49 01	<u>Halmahera</u> 0.78 S 127.44 E H = 19 30 13.8 h = normal MB=5.2 MS=4.9 D = 106.72 Az = 323.2 (NEIS)
20.	eP	A	24 00 40	<u>Turkey</u> 37.04 N 27.52 E
	e	A	01 22.5	H = 23 56 36.3 h = normal MB = 3.9
	LmH	B	09.7	D = 17.74 Az = 325.1 (NEIS)
	LmV	B	09.7	PV A 2.0s 21.4nm M = 3.9
				LmH B 13 0.3 µm
21.	ePKP2	A	02 34 56.5	<u>South of Kermadec Islands</u> 33.48 S 178.68 W H = 02 14 15.8 h = 28 km MB = 5.1 (NEIS) D = 161.2
21.	eP	A	05 27 13	<u>Albania</u> 40.12 N 19.74 E H = 05 24 22.2 h = normal MB = 4.3 D = 11.97 Az = 334.3 (NEIS)
21.	eP	A	08 41 16.5	<u>Near Coast of Chiapas, Mexico</u>
	eS	B	51 55	14.31 N 92.06 W
	LmH	B	09 25.8	H = 08 28 35.3 h = 70 km MB = 5.3
	LmV	B	25.9	D = 87.50 Az = 38.3 (NEIS)
				PV A 2.3s 61.0nm M = 5.3
				LmH B 16 1.5 µm
				LmV B 16 1.8 µm

Day	Phase		h m s	Remarks
21.	LmH	B	23 23.5	Near Coast of Nicaragua 11.27 N 86.08 W
	LmV	B	24.0	H = 22 33 38.2 h = 87 km MB = 5.2
				D = 86.22 Az = 39.2 (NEIS)
				LmH B 18s 0.8 /um
				LmV B 18 1.0 /um
22.	eP	A	03 54 57	Eastern Sea of Japan 40.66 N 137.57 E
	epP	A	56 06	H = 03 43 31.8 h = 299 km MB = 4.7
				D = 77.52 Az = 328.1 (NEIS)
				h = 303 km
22.	eP1	A	07 22 14	Southern Italy 39.36 N 15.44 E
	eP2	A	22 18	H = 07 19 34.1 h = 271 km MB = 4.7
				D = 11.60 Az = 347.8 (NEIS)
				P1V A 1.0s 19.7nm M = 4.2
				P2V A 1.9 129.0nm 4.7
22.	eP	A	13 29 47.5	Kamchatka 55.65 N 160.47 E
	epP	A	30 27	H = 13 18 46.1 h = 165 km MB = 4.8
				D = 70.95 Az = 339.6 (NEIS)
				h = 165 km
				PV A 0.9s 13.6nm M = 4.7
				pPV A traces
22.	ePg	A	15 36 47.5	Northern Italy 44.8 N 8.9 E
	eSg	A	38.07.5	H = 15 34 49 h = 0 km (ISC)
				D = 6.0
22.	e	A	16 50 04	Austria
	ei(Sg)	A	53 32.8	Explosion (VIE)
22.	ePKIKP	A	18 30 43.5	Central Chile 35.80 S 71.49 W
				H = 18 12 15.9 h = 86 km MB = 5.4
				D = 112.77 Az = 43.3 (NEIS)
				PKIKPV A 1.2s 12.2nm

Day	Phase		h m s	Remarks
23.	ePKHKP	A	00 48 02	Tonga Islands 20.72 S 174.26 W
	eX	A	48 12	H = 00 28 13.3 h = normal MB=5.0 MS=5.0
				D = 149.78 Az = 352.6 (NEIS)
				PKHKPV A 1.5s 40.2nm
				XV A 1.4 51.2nm
23.	-iPKP	AB	07 04 48.5	Tonga Islands 16.62 S 173.56 W
	eX	A	04 56	H = 06 45 12.6 h = 46 km MB=5.5 MS=4.3
	eY	AB	05 13.5	D = 145.81 Az = 354.1 (NEIS)
				PKPV A 1.6s 165.0nm
				XV A 1.2 48.8nm
				YV A 1.7 78.8nm
23.	eP	A	07 19 05	Kirgiz SSR 39.19 N 72.32 E
	e	A	19 11.5	H = 07 11 05.2 h = normal MB = 5.0
	ePcP	A	20 59	D = 43.33 Az = 306.0 (NEIS)
	LmH	B	36.0	LmH B 19s 0.6/um M = 4.5
	LmV	B	39.7	LmV B 12 0.3/um 4.5
23.	eP	A	07 44 55	Tadzhik-Sinkiang Border Region
				39.47 N 73.29 E
				H = 07 36 50.8 h = 42 km MB = 4.7
				D = 43.78 Az = 305.9 (NEIS)
				PV A 1.2s 12.2nm M = 4.5
23.	+iPKP	AB	11 18 04	New Hebrides Islands 19.53 S 169.27 E
	epPKP	AB	18 41	H = 10 58 47.5 h = 162 km MB = 5.6
				D = 144.26 Az = 335.5 (NEIS)
				h = 139 km
				PKPV A 1.4s 181.5nm
23.	eP	A	15 06 55.5	Off East Coast of Honshu, Japan
				36.63 N 142.21 E
				H = 14 54 30.6 h = 16 km MB = 4.7
				D = 82.83 Az = 330.8 (NEIS)

Day	Phase		h m s	Remarks
23.	eP	A	22 02 42	Kurile Islands 49.38 N 155.74 E
	e	A	03 07.5	H = 21 50 57.3 h = normal MB = 4.5
				D = 75.73 Az = 337.4 (NEIS)
24.	ePn	A	00 24 07	Svabian Jura Region (FRG)
	ePg	A	24 17	48.53 N 9.03 E
	eSn	A	24 38.5	H = 00 23 24.6 h = 31 km
	eSg	A	24 53	D = 2.71 Az = 37.5 (NEIS)
24.	ePKP	AB	08 47 35	Kermadec Islands Region
	ePKP2	A	48 09	31.25 S 177.80 W
	e	A	48 23	H = 08 27 35.9 h = normal MB=5.4 MS=5.8
	e	B	58 38	D = 159.39 Az = 342.8 (NEIS)
	LmH	B	10 09.3	PKIKPV A 2.3s 48.7nm
	LmV	B	11.5	PKP2V A 1.5 35.2nm
				LmH B 17.5 1.6 µm M = 5.8
				LmV B 17 1.5 µm 5.9
24.	PKHKP	A	14 33 53	Tonga Islands 18,29 S 174.56 W
				H = 14 14 11.2 h = normal MB = 5.2
				D = 147.34 Az = 352.7 (NEIS)
				PKHKPV A 1.4s 14.0nm
24.	LmV	B	21 26.8	South of Marianas 12.98 N 144.89 E
	LmH	B	28.8	H = 20 23 00.9 h = 77 km MB = 5.3 (ISC)
				D = 104.7
				LmH B 18s 0.3 µm
				LmV B 19 0.4 µm
25.	ePn	A	01 03 12	Yugoslavia 45.37 N 16.11 E
	eSn	A	04 17	H = 01 01 37.6 h = normal
	eSg	A	04 49.5	D = 6.08 Az = 331.9 (NEIS)
25.	ePKHKP	A	03 02 26	South of Fiji Islands 22.34 S 179.52 W
				H = 02 43 38.0 h = 568 km MB = 4.1
				D = 150.43 Az = 345.6 (NEIS)
				traces

Day	Phase		h m s	Remarks
25.	ePKIKP	A	17 36 38.5	**New Britain Region** 6.08 S 153.10 E
	ePP	AB	38 29	H = 17 17 38.9 h = 33 km MB=5.5 MS=5.3
	e	A	39 16	D = 125.21 Az = 331.0 (NEIS)
	e	B	44 18	PPV A 2.0s 77.0nm M = 5.7
	eSKKS	B	45 30	PPV B 8 0.8 μm 6.1
	LmH	B	18 27.8	LmH B 18 0.4 μm 5.1
	LmV	B	35.9	LmV B 18 0.5 μm 5.2
25.	ePn	A	19 11 49.5	**Federal Republic of Germany**
	ePg	A	12 01.5	50.98 N 6.41 E
	e(Sn)	A	12 25	H = 19 10 58.3 h = 8 km
	eSg	A	12 37	D = 3.32 Az = 93.8 (NEIS)
26.	eSg	A	01 12 28	**Federal Republic of Germany**
				51.2 N 6.6 E
				H = 01 10 47 (BCIS)
				D = 3.20
26.	ePn	A	01 16 40.5	**Switzerland** 46.9 N 9.8 E
	ePg	A	16 58	H = 01 15 41 (BCIS)
	eSn	A	17 29	D = 3.95
	eSg	A	17 50.5	
26.	ePP	A	13 19 58.5	**Ceram** 3.59 S 128.88 E
	LmV	B	14 18.3	H = 13 01 02.5 h = 25 km
	LmH	B	19.3	MB = 5.6 MS = 5.4 (NEIS)
				D = 109.8
				LmH B 16.5s 0.8 μm M = 5.4
				LmV B 16 1.3 μm 5.6
26.	ePKHKP	A	20 29 32	**Tonga Islands** 20.55 S 174.36 W
				H = 20 09 43.4 h = normal MB = 4.5
				D = 149.60 Az = 352.5 (NEIS)
				PKHKPV A 1.1s 10.1nm
26.	eP	A	20 49 18	**Fox Islands, Aleutian Is.**
	e	A	49 23	51.06 N 170.46 W
	e	A	49 34	H = 20 37 17.8 h = normal MB = 4.9

Day	Phase		h m s	Remarks
cont. 26.	LmH	B	21 13.0	D = 78.66 Az = 358.7 (NEIS)
	LmV	B	15.9	LmH B 16s 0.3/um M = 4.8
				LmV B 16 0.4/um 4.9
27.	epP	A	01 08 15	<u>Kurile Islands</u> 42.97 N 147.06 E
				H = 00 56 03.4 h = 44 km MB = 4.9
				D = 79.00 Az = 332.9 (NEIS)
27.	eP	AB	04 38 09	<u>Komandorsky Islands Region</u>
	ePP	B	40 50	55.50 N 166.37 E
	LmH	B	05 10.5	H = 04 26 47.0 h = normal MB=5.3 MS=5.4
	LmV	B	16.3	D = 72.15 Az = 343.4 (NEIS)
				PV A 2.3s 85.4nm M = 5.3
				PV B 6 0.7/um 5.8
				PPV B 6.5 0.7/um 6.0
				LmH B 20 2.0/um 5.4
				LmV B 16 1.9/um 5.5
27.	eP	A	10 13 40	<u>Tibet</u> 30.30 N 94.89 E
				H = 10 03 13.0 h = normal MB = 4.7
				D = 63.16 Az = 314.9 (NEIS)
				traces
27.	LmH	B	12 33.3	<u>Gulf of California</u> 27.9 N 111.07 W
	LmV	B	38.0	H = 11 44 58 h = 38 km MB = 4.4 (ISC)
				D = 86.8
27.	ePKP	A	13 44 44	<u>Tonga Islands</u> 16.36 S 175.32 W
	epPKP	A	46 04	H = 13 25 44.2 h = 334 km MB = 4.9
				D = 145.35 Az = 352.2 (NEIS)
				h = 325 km
				PKPV A 1.1s 24.2nm
				pPKPV A traces
27.	e(Sg)	A	18 14 10	

Day	Phase		h m s	Remarks
28.	eP	A	04 18 21.5	Southern Sinkiang Province, China
	epP	A	18 40.5	38.83 N 75.24 E
				H = 04 10 06.1 h = 51 km MB = 5.0
				D = 45.38 Az = 306.7 (NEIS)
				h = 76 km
				PV A 1.2s 20.3nm M = 4.9
28.	eP	A	08 13 20	Near Coast of Guerrero, Mexico
				17.86 N 101.06 W
				H = 08 00 12.1 h = normal MB = 4.5
				D = 89.91 Az = 36.0 (NEIS)
28.	+eiP	AB	11 46 54	Kurile Islands 46.29 N 153.33 E
	ePP	B	49 50	H = 11 34 59.7 h = 52 km MB = 5.9
	eS	B	56 44	D = 77.93 Az = 336.2 (NEIS)
	eiPS	C	57 35	PV A 2.1s 1390.0nm M = 6.7
	eiSS	C	12 02 10	PV B 11 9.4 μm 6.8
	LmH	B	22.9	PPV B 12 4.2 μm 6.5
	LmV	B	24.2	SH B 11.5 5.9 μm 6.6
				LmH B 15.3 48.0 μm 6.9
				LmV B 19 33.9 μm 6.7
28.	eP	A	11 52 42	Kurile Islands Region 46 N 153 E
				H = 11 40 45 MB = 4.9 (NORSAR)
				D = 78.1
				PV A 1.8s 60.9nm M = 5.3
28.	eP	A	12 04 44.5	Kurile Islands 46.33 N 153.43 E
				H = 11 52 51.9 h = 60 km MB = 5.0
				D = 77.92 Az = 336.3 (NEIS)
				PV A (1.0)s 15.7nm M = 5.0
28.	eP	A	12 11 38.5	Kurile Islands 46.58 N 153.32 E
				H = 11 59 46.0 h = 49 km MB = 4.8
				D = 77.67 Az = 336.2 (NEIS)
				PV A 1.1s 16.1nm M = 5.0

Day	Phase		h m s	Remarks
28.	eP	A	12 19 45	Kurile Islands 46.18 N 153.26 E
	epP	A	19 59	H = 12 07 50.0 h = 46 km MB = 5.3
				D = 78.02 Az = 336.2 (NEIS)
				h = 50 km
				PV A 1.3s 52.4nm $\dot{M}$ = 5.4
				pPV A 2.0 179.5nm
28.	eP	A	12 31 35	Kurile Islands 46.24 N 153.35 E
				H = 12 19 39.0 h = normal MB = 4.8
				D = 77.98 Az = 336.2 (NEIS)
				PV A 1.2s 18.3nm M = 5.0
28.	eP	A	13 16 11.5	Kurile Islands 46.35 N 153.28 E
				H = 13 04 17.5 h = 44 km MB = 4.8
				D = 77.86 Az = 336.2 (NEIS)
				PV A 1.2s 16.3nm M = 4.9
28.	e	A	13 42 50	Kurile Islands 46.16 N 153.57 E
				H = 13 30 50 h = 19 km MB = 4.2
				D = 78.12 Az = 336 (ISC)
28.	eP	A	13 43 31.5	Kurile Islands 46.27 N 153.51 E
				H = 13 31 39.3 h = 66 km MB = 5.3
				D = 78.00 Az = 336.3 (NEIS)
				PV A 2.6s 277.0nm M = 5.8
28.	eP	A	13 53 33	Kurile Islands 46.32 N 153.32 E
	e	A	53 38	H = 13 41 38.8 h = 46 km MB = 5.4
				D = 77.90 Az = 336.2 (NEIS)
				PV A 2.0s 128.0nm M = 5.6
28.	eP	A	15 22 29	Kurile Islands 46.22 N 153.14 E
	epP	A	22 40	H = 15 10 34.7 h = 46 km MB = 4.5
				D = 77.94 Az = 336.1 (NEIS)
				h = 39 km

Day	Phase		h m s	Remarks
28.	eP	A	16 34 52	Kurile Islands Region 45.95 N 153.10 E
	epP	A	35 04	H = 16 22 55.3 h = normal MB = 4.7
				D = 78.18 Az = 336.1 (NEIS)
				h = 44 km
28.	eP	A	16 39 28	Kurile Islands 46.09 N 153.28 E
				H = 16 27 32.6 h = 52 km MB = 5.1
				D = 78.10 Az = 336.2 (NEIS)
				PV A 1.9s 45.5nm M = 5.2
28.	eP	A	16 45 50	Kurile Islands 46.25 N 153.23 E
	LmH	B	17 19.6	H = 16 33 55.6 h = 49 km MB=4.9 MS=5.0
	LmV	B	25.5	D = 77.94 Az = 336.2 (NEIS)
				PV A 1.1s 24.2nm M = 5.1
				LmH B 16.5 1,1 µm 5.3
				LmV B 15 1.0 µm 5.3
28.	eP	A	17 08 28	Kurile Islands 46.20 N 152.98 E
				H = 16 56 32.4 h = 43 km MB = 4.3
				D = 77.92 Az = 336.0 (NEIS)
28.	eP	A	17 17 36	Kurile Islands 46.27 N 153.11 E
				H = 17 05 40.0 h = 42 km MB = 4.8
				D = 77.88 Az = 336.1 (NEIS)
28.	eP	A	18 12 39.5	Kurile Islands 46.42 N 153.36 E
	LmH	B	48.8	H = 18 00 45.4 h = 46 km MB=4.9 MS=4.6
	LmV	B	56.0	D = 77.82 Az = 336.2 (NEIS)
				PV A 2.4s 96.6nm M = 5.4
				LmH B 15.5 0.7 µm 5.1
				LmV B 15 0.5 µm 5.0
28.	eP	A	18 31 16	Kurile Islands 46.00 N 153.38 E
				H = 18 19 19.1 h = 40 km MB = 4.2
				D = 78.21 Az = 336.3 (NEIS)
				traces

Day	Phase		h m s	Remarks
28.	ePKP2	A	20 35 44	Kermadec Islands Region
	LmH	B	21 57.3	31.32 S 177.83 W
	LmV	B	22 01.3	H = 20 15 08.2 h = 14 km
				MB = 4.9 MS = 5.1 (NEIS)
				D = 159.4
28.	eP	A	22 03 10	Talaud Islands 2.75 N 125.67 E
				H = 21 49 21.6 h = 100 km MB = 5.6
				D = 102.86 Az = 323.4 (NEIS)
				PV A 1.0s 15.7nm M = 5.8
29.	eP	A	01 21 17.5	Turkey 30.0 N 41.1 E
	LmH	B	32.5	H = 01 16 06 (BCIS)
	LmV	B	32.5	D = 23.72
				LmH B 16s 0.3/um M = 3.9
				LmV B 16 0.35/um 4.1
29.	eP	A	01 26 41	Kurile Islands 46.88 N 152.81 E
				H = 01 14 51.0 h = 54 km MB = 5.0
				D = 77.25 Az = 335.9 (NEIS)
29.	eP	A	02 25 42.5	Kurile Islands 46.24 N 153.19 E
	epP	A	25 53	H = 02 13 47.2 h = 42 km MB=4.8 MS=4.3
				D = 77.94 Az = 336.1 (NEIS)
				h = 35 km
				PV A 0.9s 11.7nm M = 4.9
29.	eP	AB	03 27 12	Kurile Islands 46.22 N 153.06 E
	e	B	27 48	H = 03 15 16.7 h = 38 km MB=5.7 MS=5.8
	eS	B	37 03	D = 77.92 Az = 336.1 (NEIS)
	LmH	B	04 06.6	PV A 2.0s 368.0nm M = 6.1
	LmV	B	06.6	PV B 10 4.0/um 6.5
				LmH B 17.5 14.7/um 6.4
				LmV B 18 6.4/um 6.4
29.	eP	A	04 11 27	Kurile Islands 46.21 N 153.22 E
				H = 03 59 28.8 h = normal MB = 4.3
				D = 77.97 Az = 336.2 (NEIS)

Day	Phase		h m s	Remarks
29.	+iP1	AB	07 28 22.5	<u>Kurile Islands</u> 46.12 N 153.09 E
	eP2	A	28 28	H = 07 16 26.1 h = normal MB=5.9 MS=6.2
	e	B	28 56	D = 78.02 Az = 336.1 (NEIS)
	eS	B	38 13	P1V A 1.3s 201.0nm M = 6.1
	LmV	B	08 05.9	P2V A 2.0 423.0nm 6.2
	LmH	B	06.1	P1V B 10 6.3 µm 6.7
				LmH B 17 27.8 µm 6.6
				LmV B 19.5 21.4 µm 6.5
29.	eP	A	09 40 44	<u>Kurile Islands</u> 46.33 N 153.33 E
				H = 09 28 48.4 h = normal MB=4.9 MS=5.5
				D = 77.89 Az = 336.2 (NEIS)
29.	ePKP	A	12 11 50	<u>Tonga Islands</u> 17.96 S 175.23 W
	epPKP	A	12 57.5	H = 11 52 36.9 h = 260 km MB = 4.9
				D = 146.93 Az = 352.0 (NEIS)
				h = 272 km
				PKPV A 1.3s 17.5nm
29.	eP	A	14 33 12.5	<u>Kurile Islands</u> 46.15 N 152.91 E
	LmH	B	15 09.8	H = 14 21 16.2 h = 36 km MB = 4.9
	LmV	B	10.8	D = 77.94 Az = 336.0 (NEIS)
				PV A 2.0s 42.7nm M = 5.1
				LmH B 17 0.35 µm 5.2
				LmV B 18 0.5 µm 5.4
29.	eP	A	19 21 54.5	<u>Kurile Islands</u> 45.99 N 153.18 E
				H = 19 09 57.7 h = 42 km MB = 4.6
				D = 78.16 Az = 336.1 (NEIS)
29.	eP	A	20 02 38	<u>Kurile Islands</u> 45.98 N 153.01 E
				H = 19 50 41.4 h = normal MB = 4.9
				D = 78.12 Az = 336.0 (NEIS)
29.	ePKIKP	AB	22 32 47.5	<u>Fiji Islands Region</u> 17.90 S 178.52 W
	ePKHKP	A	32 49	H = 22 14 12.8 h = 586 km MB = 5.4
	ePKP2	A	32 51.5	D = 146.34 Az = 348.3 (NEIS)
	e	A	33 57	PKHKPV A 1.4s 86.0nm
	e	A	35 05.5	

Day	Phase		h m s	Remarks
30.	+iP	AB	05 20 29.0	Hindu Kush Region 36.35 N 70.76 E
	i	A	20 34.0	H = 05 12 40.6 h = 211 km MB = 6.5
	epP	B	21 16	D = 44.05 Az = 308.2 (NEIS)
	eiPP	B	22 18	PV A 1.4s 2260.0nm M = 6.4
	eiS	B	26 44	PV B 12 21.1/um 6.5
	eisS	B	28 04	SH B 18 57.2/um 7.0
	eSS	B	29 55	LmH B 19 88.6/um
	LmV	B	31.1	LmV B 12 45.5/um
	LmH	B	31.3	
30.	ePKP	A	10 47 02	Tonga Islands 15.94 S 173.10 W
				H = 10 27 26.5 h = 33 km MB = 4.8
				D = 145.17 Az = 354.7 (NEIS)
				PKPV A 0.7s 11.5nm
30.	eP	A	11 49 38	West Pakistan 35.54 N 71.51 E
	e	A	50 15	H = 11 41 27.4 h = 70 km MB = 5.3
	LmH	B	12 08.3	D = 45.03 Az = 308.9 (NEIS)
	LmV	B	12.6	PV A 1.6s 44.0nm M = 5.0
30.	ePKHKP	A	22 01 53	Fiji Islands Region 17.84 S 178.61 W
	epPKP	A	04 19.5	H = 21 43 18.8 h = 613 km MB = 5.3
				D = 146.26 Az = 348.3 (NEIS)
30.	eP	A	22 51 39	Kurile Islands 46.22 N 153.17 E
	epP	A	51 52.5	H = 22 39 44.5 h = 42 km MB=5.0 MS=4.7
				D = 77.95 Az = 336.1 (NEIS)
				h = 48 km
				PV A 1.3s 30.6nm M = 5.2
31.	ePKP2	A	22 51 37.5	Kermadec Islands Region 28.40 S 176.65 W
				H = 22 31 18.2 h = 58 km MB = 4.8
				D = 156.92 Az = 346.5 (NEIS)

Day	Phase		h m s	Remarks
1.	e	A	01 45 46	Solomon Islands 8.57 S 157.56 E
	LmV	B	45.5	H = 01 26 30.1 h = 30 km MB=4.8 MS=5.1
	LmH	B	45.6	D = 129.48 Az = 332.5 (NEIS)
1.	eP	A	05 17 44	Kodiak Island Region 56.72 N 152.10 W
				H = 05 06 19.5 h = 24 km MB = 4.6
				D = 72.15 Az = 10.8 (NEIS)
1.	ePKP	AB	05 18 48	Loyalty Islands Region 22.35 S 170.64 E
	eX	AB	19 29.5	H = 04 59 08.3 h = 52 km MB = 5.6
	eSKKS	B	28 56	D = 147.34 Az = 335.0 (NEIS)
	LmH	B	06 43.0	PKPV A 2.0s 94.0nm
	LmV	B	46.5	PKPV B 8 1.0 µm
				XV A 1.6 115.4nm
				XV B 8 1.6 µm
				LmH B 17.5 7.5 µm
				LmV B 17 10.6 µm
1.	eP	AB	06 07 02	Kodiak Island Region 56.67 N 152.11 W
	ePP	A	09 33.5	H = 05 55 38.2 h = normal MB=5.7 MS=6.3
				D = 72.21 Az = 10.8 (NEIS)
				PV A 1.3s 162.0nm M = 5.9
				PPV A 1.3 43.7nm 5.5
1.	eP	A	06 18 44	Kodiak Island Region 56.59 N 152.39 W
				H = 06 07 17.0 h = 15 km MB = 4.6
				D = 72.31 Az = 10.6 (NEIS)
1.	eP	A	06 28 28	Kodiak Island Region 56.65 N 152.07 W
				H = 06 17 02.3 h = 19 km MB = 4.5
				D = 72.23 Az = 10.8 (NEIS)
1.	eP	A	06 59 42	Kodiak Island Region 56.71 N 152.20 W
				H = 06 48 16.3 h = 16 km MB = 4.4
				D = 72.17 Az = 10.7 (NEIS)
1.	eP	AB	08 11 21	Kodiak Island Region 56.63 N 152.27 W
	eS	B	20 48	H = 07 59 56.9 h = normal MB=5.2 MS=6.0

Day	Phase		h m s	Remarks
cont. 1.	LmH	B	08 47.7	D = 72.26 Az = 10.7 (NEIS)
	LmV	B	51.0	PV A 1.8s 60.9nm M = 5.4
				PV B 5.5 0.45/um 5.8
				LmH B 17 1.3/um 5.3
				LmV B 16 1.4/um 5.4
1.	eP	A	09 47 32	Zambia 16.65 S 28.00 E
				H = 09 36 27.0 h = 14 km MB = 5.1
				D = 68.54 Az = 348.9 (NEIS)
				PV A 1.4s 20.93nm M = 4.9
1.	e(Sg)	A	15 47 23.5	
1.	LmH	B	21 07.5	South-East Indian Ridge 41.96 S 88.4 E
	LmV	B	12.4	H = 19 51 31.3 h = 33 km MB = 5.0 (ISC)
				or
				South-East Indian Ridge 41.93 S 88.38 E
				H = 19 57 30.4 h = 33 km MB = 4.9 (ISC)
				D = 113.8
				LmH B 18s 0.5/um M = 5.2
				LmV B 17 0.6/um 5.3
1.	eP	AB	22 51 03	Kurile Islands 49.76 N 155.97 E
	e	A	51 37	H = 22 39 21.0 h = 41 km MB=5.3 MS=5.2
	LmH	B	23 26.5	D = 75.44 Az = 337.5 (NEIS)
	LmV	B	28.7	LmH B 20s 2.2/um M = 5.5
				LmV B 18 1.4/um 5.3
2.	e(P)	A	08 30 41	Iran 30.46 N 50.59 E
	LmH	B	48.8	H = 08 23 44.0 h = 44 km MB = 4.8
	LmV	B	48.8	D = 35.28 Az = 316.1 (NEIS)
				LmH B 15s 0.4/um M = 4.3
				LmV B 14.5 0.6/um 4.6
2.	eP	A	10 03 04.5	South of Honshu, Japan 33.38 N 139.37 E
				H = 09 50 33.9 h = 38 km MB=4.8 MS=4.5
				D = 84.48 Az = 329.6 (NEIS)

Day	Phase		h m s	Remarks
2.	eP	A	10 34 32	Crete 35.57 N 26.71 E H = 10 30 14.4 h = 50 km MB = 4.3 D = 18.61 Az = 328.7 (NEIS)
2.	ePKP2	A	10 42 04	Kermadec Islands Region 31.47 S 177.82 W H = 10 21 30 h = 29km D = 159.60 Az = 343 (ISC)
2.	eP LmH LmV	A B B	14 45 59 15 23.2 28.8	South of Honshu, Japan 33.35 N 139.41 E H = 14 33 26.6 h = 24 km MB=5.1 MS=4.5 D = 84.53 Az = 329.6 (NEIS) LmH B 15s 0.4 µm M = 4.9 LmV B 14 0.3 µm 4.9
2.	eP	A	16 17 56.5	South of Honshu, Japan 33.40 N 139.28 E H = 16 05 25.1 h = 34 km MB = 4.8 D = 84.43 Az = 329.6 (NEIS)
2.	eP LmH LmV	A B B	16 46 25 59.0 17 01.0	North Atlantic Ridge 32.71 N 39.76 W H = 16 38 40.0 h = normal MB = 4.1 D = 41.43 Az = 48.7 (NEIS) PV A traces LmH B 14s 0.3 µm M = 4.4 LmV B 16 0.5 µm 4.5
2.	ePKP epPKP	A A	18 18 10 19 20	Tonga Islands 17.30 S 175.28 W H = 17 59 00.9 h = 281 km MB = 4.9 D = 146.28 Az = 352.1 (NEIS) h = 282 km PKPV A traces
3.	eP LmH LmV	A B B	04 17 16 37.4 41.0	Kashmir-Tibet Border Region 35.45 N 80.64 E H = 04 08 13.8 h = 20 km MB = 5.0 D = 50.85 Az = 310.0 (NEIS) LmH B 16s 0.8 µm M = 4.8 LmV B 15 0.7 µm 4.9

Day	Phase		h m s	Remarks
3.	e(PKP)	A	11 54 44.5	New Hebrides Islands 19.01 S 169.54 E
				H = 11 35 42.6 h = 275 km MB = 4.8
				D = 143.89 Az = 336.0 (NEIS)
3.	eP	A	12 43 56.5	Kurile Islands 46.64 N 152.79 E
				H = 12 32 05.3 h = 56 km MB = 4.8
				D = 77.46 Az = 335.9 (NEIS)
3.	eP	A	13 57 33.5	Northern Sumatra 0.42 N 98.66 E
				H = 13 44 45.9 h = normal MB = 5.0 (NEIS)
				D = 87.80
3.	eP	A	14 06 07	Kurile Islands 45.67 N 149.62 E
				H = 13 54 26.1 h = 142 km MB = 4.4
				D = 77.40 Az = 334.1 (NEIS)
				PV A traces
3.	e	A	15 00 04.5	Tonga 15.6 S 173.9 W
				H = 14 40 09.7 h = 0 km MB = 4.5
				D = 144.77 Az = 354 (ISC)
3.	eP1	AB	18 28 50.5	Honshu, Japan 36.00 N 139.82 E
	+iP2	A	28 53.5	H = 18 16 34.0 h = 58 km MB = 5.6
	ePP	AB	32 02.5	D = 82.42 Az = 329.7 (NEIS)
	eS	B	39 04	P1V A traces
	eSS	C	44 24	P1V B 5s 1.0/um M = 6.1
	LmH	B	19 09.6	P2V A 1.8 236.0nm 5.9
	LmV	B	09.7	PPV A 1.5 55.3nm 5.7
				LmH B 16.2 2.0/um 5.6
				LmV B 15.5 2.4/um 5.6
4.	eP	A	00 45 51	Kurile Islands 46.09 N 153.44 E
				H = 00 33 56.0 h = 60 km MB = 4.5
				D = 78.15 Az = 336.3 (NEIS)
4.	LmH	B	11 20.3	LmH B 20s 0.4/um
	LmV	B	21.0	LmV B 20 0.5/um

Day	Phase		h m s	Remarks
4.	+iP	A	15 11 38.7	<u>Eastern Caucasus</u> 42.34 N 45.93 E
	eS	B	16 16	H = 15 06 17.1 h = normal MB=5.4 MS=5.0
	LmH	B	22.4	D = 24.86 Az = 301.4 (NEIS)
	LmV	B	23.1	PV A 1.2s 187.0nm M = 5.5
				SH B 7.5 1.6/um 5.5
				LmH B 14 2.0/um 4.8
				LmV B 12 1.4/um 4.8
4.	eP	A	21 47 41	<u>Kodiak Island Region</u> 56.76 N 152.01 W
				H = 21 36 16.7 h = 40 km MB = 4.3 (NEIS)
				D = 72.11
				traces
5.	eP	A	13 27 06	<u>Southern Iran</u> 27.98 N 53.55 E
	LmH	B	47.8	H = 13 19 39.5 h = 11 km MB = 5.3
	LmV	B	47.8	D = 38.85 Az = 317.3 (NEIS)
				PV A 1.1s 16.1nm M = 4.7
5.	ePn	A	18 39 27	<u>Northern Italy</u> 44.4 N 7.5 E
	eSn	A	40 36	H = 18 37 46 (BCIS)
	eSg	A	41 35	D = 6.87
				PnV A traces
6.	eP	A	02 48 33.5	<u>Southern Alaska</u> 60.25 N 153.32 W
	ipP	A	49 06.5	H = 02 37 42.3 h = 136 km MB = 5.0
				D = 68.80 Az = 10.2 (NEIS)
				h = 133 km
				pPV A 1.4s 32.6nm
6.	eP1	A	11 17 00	<u>North of Svalbard</u> 80.0 N 3.0 E
	eP2	A	17 08	H = 11 10 52 (BCIS)
	LmV	C	34	D = 29.6
	LmH	C	37	P1V A 1.1s 20.2nm M = 4.8
				P2V A 1.8 47.3nm 5.0
				LmV C 14 0.3/um 4.2
6.	eP	A	13 11 23	<u>Kodiak Island Region</u> 56.55 N 152.47 W
				H = 12 59 55.6 h = 16 km MB=4.5 MS=4.7
				D = 72.36 Az = 10.6 (NEIS)

Day	Phase		h m s	Remarks
6.	eP	A	17 00 57	Kurile Islands 46.76 N 150.42 E H = 16 49 22.2 h = 166 km MB = 5.2 D = 76.67 Az = 334.5 (NEIS) PV A 1.6s 49.5nm M = 5.0
6.	eP	A	18 16 49	Kodiak Island Region 56.72 N 152.41 W H = 18 05 25.3 h = normal MB = 4.6 D = 72.19 Az = 10.6 (NEIS)
6.	ePKIKP ePKHKP eipPKHKP LmV LmH	AB A A B B	18 57 56 58 02 58 17.5 20 14.3 15.2	Tonga Islands 21.76 S 175.16 W H = 18 38 13.1 h = 48 km MB=5.7 MS=5.7 D = 150.68 Az = 351.2 (NEIS) h = 57 km PKIKPV A 1.3s 17.5nm PKHKPV A 1.5 126.0nm LmH B 18.7 1.8/um M = 5.8 LmV B 18 2.0/um 6.0
6.	eP	A	20 18 03.5	Iran-USSR Border Region 37.8 N 58.4 E H = 20 11 05 h = 0 km MB = 4.4 D = 35.27 Az = 307 (ISC) PV A 1.0s 11.8nm M = 4.6
7.	eP	A	00 57 16.5	Greenland Sea 73.46 N 6.89 E H = 00 52 14.9 h = normal MB = 4.6 D = 22.99 Az = 172.3 (NEIS) PV A 1.0s 19.7nm M = 4.6
7.	eP	A	01 05 31	Greenland Sea 73.14 N 7.30 E H = 01 00 28.8 h = normal MB = 4.1 D = 22.65 Az = 172.9 (NEIS)
7.	eP	A	01 07 15	Norwegian Sea 70.5 N 9.7 E H = 01 02 44.5 h = 33 km D = 19.91 Az = 176 (ISC)
7.	eP	A	01 20 11.5	Greenland Sea 73.4 N 8.0 E H = 01 15 11 h = 33 km D = 22.87 Az = 174 (ISC)

Day	Phase		h m s	Remarks
7.	eP	A	01 52 49.5	Greenland Sea 73.30 N 6.68 E H = 01 47 48.2 h = normal MB = 4.5 D = 22.84 Az = 171.9 (NEIS) PV A 1.1s 14.1nm M = 4.4
7.	eP	A	03 58 10	Norwegian Sea 70.7 N 10.2 E H = 03 53 36 h = 33 km D = 20.14 Az = 177 (ISC)
7.	eP	A	04 01 02	Greenland Sea 73.37 N 7.1 E H = 03 55 54.5 h = 0 km D = 22.89 Az = 173 (ISC)
7.	eP	A	08 32 25.5	Kodiak Island Region 56.78 N 152.30 W H = 08 21 01.8 h = 32 km MB = 4.6 D = 72.11 Az = 10.7 (NEIS) PV A 1.0s 13.8nm M = 4.9
7.	+eP	AB	08 35 02	Kodiak Island Region 56.65 N 152.31 W
	eS	B	44 27	H = 08 23 36.8 h = normal MB=4.9 MS=5.5
	LmH	B	09 09.5	D = 72.25 Az = 10.7 (NEIS)
	LmV	B	13.8	PV A 1.0s 29.5nm M = 5.2 LmH B 18.5 1.0 μm 5.1 LmV B 18 1.2 μm 5.3
7.	eP	A	08 38 11	Kodiak Islands Region 56.7 N 152.3 W H = 08 26 42 h = 0 km MB = 5.2 D = 72.17 Az = 11 (ISC) PV A 1.4s 23.2nm
7.	ePn	A	15 40 09.5	D c. 5
	eSn	A	41 08	
	eSg	A	41 36	
7.	eP	A	19 03 35	Talaud Islands 3.07 N 125.61 E H = 18 49 54.9 h = 158 km MB = 5.4 D = 102.56 Az = 323.4 (NEIS) PV A 1.5s 20.1nm M = 5.7

Day	Phase		h m s	Remarks
7.	eP	A	21 16 41	Afghanistan-USSR Border Region
				36.16 N 71.00 E
				H = 21 08 43.3 h = 144 km MB = 4.8
				D = 44.32 Az = 308.4 (NEIS)
8.	eP	A	01 27 58.5	Greenland Sea 73.28 N 6.9 E
				H = 01 22 52.8 h = 0 km MB = 4.0
				D = 22.81 Az = 172 (ISC)
8.	eP	AB	01 30 17	Greenland Sea 73.20 N 6.19 E
	eS	B	34 26	H = 01 25 15.8 h = normal MB=5.0 MS=5.2
	LmH	B	39.8	D = 22.76 Az = 171.1 (NEIS)
	LmV	B	39.9	PV A 1.3s 50.2nm M = 4.8
				SH B 15 2.2 µm 5.3
				LmH B 15 2.8 µm 4.8
				LmV B 16 3.3 µm 5.0
8.	ePKHKP	A	10 40 50	Fiji Islands Region 19.76 S 177.81 W
				H = 10 22 06.3 h = 569 km MB = 5.0
				D = 148.29 Az = 348.6 (NEIS)
				PKHKPV A 1.6s 46.7nm
8.	eP	AB	19 10 01	Greenland Sea 73.17 N 5.86 E
	eS	B	14 12	H = 19 05 00.9 h = normal MB=4.8 MS=4.6
	LmH	B	14.4	D = 22.75 Az = 170.5 (NEIS)
				PV A 1.8s 54.1nm M = 4.7
				PV B 4.8 0.45 µm 5.2
				LmH B 12 0.35 µm 4.0
8.	eP	A	19 29 19	Taiwan Region 24.57 N 122.69 E
	eS	B	39 48	H = 19 16 45.9 h = 32 km MB=5.4 MS=6.0
	LmH	B	20 05.1	D = 83.66 Az = 323.3 (NEIS)
	LmV	B	10.1	PV A 1.4s 23.2nm M = 5.1
				SH B 15 0.9 µm 5.7
				LmH B 15.7 24.8 µm 6.7
				LmV B 14 17.1 µm 6.6

Day	Phase		h m s	Remarks
8.	eP	AB	23 29 44	Greenland Sea 73.40 N 6.49 E
	eS	B	33 54	H = 23 24 42.4 h = normal MB=4.7 MS=4.5
	LmH	B	39.3	D = 22.94 Az = 171.6 (NEIS)
	LmV	B	39.3	PV A 1.4s 27.9nm M = 4.5
				LmH B 16 0.4/um 4.0
				LmV B 15 0.5/um 4.2
9.	e	A	02 44 22	Ascension Island Region 10.46 S 13.10 W
				H = 02 33 40.2 h = normal MB = 4.7
				D = 64.55 Az = 17.1 (NEIS)
9.	eP	A	05 06 32.5	Peru-Brazil Border Region
	e	A	06 50	8.48 S 74.28 W
				H = 04 53 30.7 h = 159 km MB = 5.6
				D = 93.89 Az = 39.5 (NEIS)
9.	eP	A	20 18 49	Albania 40.15 N 19.9 E
	e	A	21 18	H = 20 16 02 h = 47 km
	LmH	B	24.5	D = 11.98 Az = 334 (ISC)
	LmV	B	24.6	LmH B 9s 0.2/um
9.	ePn	A	22 19 51	Federal Republic of Germany
	eSg	A	20 38	51.50 N 7.10 E
				H = 22 19 04.0 h = 0 km
				D = 2.97 Az = 104.9 (NEIS)
9.	eP	A	22 31 54	Albania 40.01 N 19.74 E
	e	A	34 56	H = 22 28 59.7 h = 0 km
				D = 12.06 Az = 334 (ISC)
10.	ePKHKP	A	05 38 19	Tonga Region 22.1 S 174.4 W
				H = 05 18 32 h = 76 km
				D = 151.12 Az = 352 (ISC)
10.	iPn	A	08 47 51	Přibram, (PRU)
10.	ePKP	A	10 06 15	Fiji Islands Region 18.54 S 179.47 W
	e	A	06 19.5	H = 09 47 39.2 h = 595 km MB = 5.1

Day	Phase		h m s	Remarks
cont. 10.	e	A	10 06 19.5	D = 146.77 Az = 347.1 (NEIS)
				PKPV A 1.8s 67.6 nm
10.	ePn	A	11 00 54	
	eSg	A	01 32.5	
10.	ePKIKP	AB	11 41 03.5	<u>Fiji Islands Region</u> 21.39 S 179.16 W
	eiPKHKP	A	41 09	H = 11 22 26.4 h = 602 km MB = 5.5
	ePKP2	A	41 16.5	D = 149.59 Az = 346.4 (NEIS)
	epPKP	A	43 30	PKIKPV A 1.8s 87.9nm
				PKHKPV A 1.6 368.0nm
				PKP2V A 1.6 93.4nm
10.	eP	A	12 52 31	<u>United Kingdom</u> 57.15 N 5.21 W
	eS	A	54 48	H = 12 49 38.0 h = 10 km MB = 4.3 (NEIS)
	LmH	B	57.5	D = 11.85
	LmV	B	58.9	PV A 0.8s 15.4nm M = 5.3
				LmH B 12.5 0.8/um 3.9
				LmV B 12 0.3/um
10.	epP	A	15 14 53	<u>Near East Coast of Honshu, Japan</u>
				40.14 N 142.42 E
				H = 15 02 35.0 h = 50 km MB = 4.7
				D = 79.85 Az = 330.7 (NEIS)
				pPV A 1.8s 23.6nm
10.	e	A	17 03 20	<u>Yougoslavia</u> 42 N 18 E
				H = 16 59 14 MB = 2.7 (NORSAR)
				D = 9.8
10.	ePKP	A	19 13 44.5	<u>Samoa Islands Region</u> 16.14 S 172.82 W
				H = 18 54 09.7 h = 42 km MB = 4.3
				D = 145.39 Az = 355.0 (NEIS)
10.	eP	A	19 38 35	<u>Kurile Islands</u> 46.24 N 153.19 E
				H = 19 26 36.8 h = normal MB = 4.4
				D = 77.94 Az = 336.1 (NEIS)

Day	Phase		h m s	Remarks
11.	-eP	AB	01 22 06	Tadzhik-Sinkiang Border Region
	ePP	B	24 00	39.46 N 73.83 E
	eS	B	28 40	H = 01 13 55.5 h = 9 km MB=6.4 MS=7.3
	eSS	B	32 50	D = 44.12 Az = 306.0 (NEIS)
	LmH	B	41.5	PV A 1.2s 732.0nm M = 6.4
	LmV	B	41.7	PV B 10 15.0 μm 6.8
				PPV B 12 26.2 μm 7.0
				SH B 16 45.1 μm 7.0
				LmH B 20 1794.0 μm 8.1
				LmV B 18 648.4 μm 7.7
11.	eP	A	02 37 55	Tadzhik-Sinkiang Border Region
				39.39 N 73.82 E
				H = 02 29 48.4 h = normal MB = 4.7
				D = 44.15 Az = 306.0 (NEIS)
11.	eP	A	02 45 17.5	Tadzhik-Sinkiang Border Region
				39.37 N 73.55 E
				H = 02 37 11.1 h = normal MB = 5.0
				D = 44.00 Az = 306.0 (NEIS)
11.	eP	A	04 36 54.5	Tadzhik-Sinkiang Border Region
				39.29 N 73.89 E
				H = 04 28 46.9 h = normal MB = 5.1
				D = 44.26 Az = 306.1 (NEIS)
11.	eP	A	05 20 40	Tadzhik-Sinkiang Border Region
	ePP	A	22 26.5	39.32 N 73.78 E
				H = 05 12 33.3 h = normal MB = 5.4
				D = 44.17 Az = 306.1 (NEIS)
				PV A 1.2s 44.7nm M = 5.2
				PPV A 1.7 60.6nm 5.2
				LmH B 18.5 4.3 μm 5.4
				LmV B 14 1.9 μm 5.2
11.	eP1	A	05 27 40	Tadzhik-Sinkiang Border Region
	eP2	A	27 44	39.36 N 73.84 E

Day	Phase		h m s	Remarks
cont. 11.	eP3	A	05 27 46.5	H = 05 19 33.2 h = 32 km MB = 5.2
				D = 44.19 Az = 306.1 (NEIS)
				P3V A 1.2s 24.4nm M = 4.9
11.	eP	A	05 32 00	<u>Tadzhik-Sinkiang Border Region</u>
	ePP	A	33 45	39.42 N 73.77 E
	LmH	B	49.3	H = 05 23 52.5 h = 27 km MB = 5.6
	LmV	B	52.4	D = 44.11 Az = 306.0 (NEIS)
				PV A 1.1s 88.7nm M = 5.5
				PPV A 1.8 74.3nm 5.3
				LmH B 14 5.0/um 5.6
				LmV B 12 2.4/um 5.4
11.	eP	A	07 10 18	<u>Tadzhik-Sinkiang Border Region</u>
	LmH	B	29.3	39.35 N 73.85 E
	LmV	B	30.1	H = 07 02 08.5 h = normal MB=5.2 MS=5.4
				D = 44.20 Az = 306.1 (NEIS)
				PV A 0.9s 19.5nm M = 4.9
				LmH B 17 8.5/um 5.7
				LmV B 14 4.0/um 5.6
11.	eP	A	08 11 05	<u>Tadzhik-Sinkiang Border Region</u>
	ePP	A	12 53	39.29 N 73.92 E
				H = 08 02 54.0 h = 12 km MB = 5.1
				D = 44.28 Az = 306.2 (NEIS)
11.	eP	A	09 17 10.5	<u>Tadzhik-Sinkiang Border Region</u>
	e	A	21 45	39.24 N 73.86 E
				H = 09 08 58.5 h = 29 km MB = 5.1
				D = 44.27 Az = 306.2 (NEIS)
				PV A 1.8s 33.8nm M = 4.9
11.	LmH	B	12 33.7	<u>Tadzhikistan-Sinkiang Border Region</u>
				39.33 N 73.79 E
				H = 12 06 25.1 h = 52 km MB = 4.4 (ISC)
				D = 44.3
				LmH B 16s 0.5/um

Day	Phase		h m s	Remarks
11.	LmH	B	13 48.9	<u>Tadzhikistan-Sinkiang Border Region</u>
	LmV	B	49.4	39.18 N 73.78 E
				H = 13 21 20 h = 53 km MB = 4.2 (ISC)
				D = 44.3
				LmH B 16s 0.6/um
				LmV B 14 0.3/um
11.	eP1	AB	20 13 36	<u>Tadzhik-Sinkiang Border Region</u>
	iP2	A	13 38.2	39.47 N 73.65 E
	ePP	A	15 22	H = 20 05 30.1 h = normal MB=5.8 MS=5.7
	e	B	15 40	D = 44.00 Az = 306.0 (NEIS)
	eS	B	20 10	P1V A 1.2s 67.1nm M = 5.3
	eSS	B	23 20	P2V A 1.8 270.0nm 5.7
	LmH	B	32.6	LmH B 17 21.4/um 6.1
	LmV	B	33.4	LmV B 13 11.7/um 6.1
11.	ePKP	A	21 28 45	<u>Fiji Islands Region</u> 20.2 S 177.7 W
				H = 21 09 02.6 h = 0 km MB = 4.8
				D = 148.77 Az = 349 (ISC)
11.	eP	AB	21 29 43	<u>Tadzhik-Sinkiang Border Region</u>
	ePP	AB	31 32.5	39.47 N 73.63 E
	eS	B	36 16	H = 21 21 33.8 h = 9 km MB=5.9 MS=6.1
	eSS	B	39 28	D = 43.99 Az = 306.0 (NEIS)
	LmH	B	48.9	PV A 1.9s 37.2nm M = 5.9
	LmV	B	49.9	PPV A 2.1 498.3nm 6.0
				SH B 11 2.0/um 5.8
				LmH B 16 33.2/um 6.4
				LmV B 12 30.4/um 6.5
11.	eP	A	21 58 14	<u>Tadzhik-Sinkiang Border Region</u>
				39.44 N 73.61 E
				H = 21 50 07.8 h = normal. MB = 4.8
				D = 43.99 Az = 306.0 (NEIS)
11.	eP	A	22 04 37	<u>Tadzhik-Sinkiang Border Region</u>
				39.59 N 73.55 E
				H = 21 56 29.9 h = 33 km MB = 4.8
				D = 43.87 Az = 305.8 (NEIS)

Day	Phase		h m s	Remarks
11.	eP	A	22 18 36	Tadzhik-Sinkiang Border Region
				39.42 N 73.71 E
				H = 22 10 27.1 h = normal MB = 4.6
				D = 44.07 Az = 306.0 (NEIS)
11.	eP	A	23 27 04	Tadzhik-Sinkiang Border Region
				39.46 N 73.60 E
				H = 23 18 58.3 h = normal MB = 5.1
				D = 43.98 Az = 306.0 (NEIS)
				PV A 1.2s 20.3nm M = 4.8
12.	ePKP	AB	03 12 12	Fiji Islands Region 15.98 S 179.41 W
	LmH	B	04 20.8	H = 02 52 42.4 h = 70 km MB = 5.7
	LmV	B	24.2	D = 144.30 Az = 348.0 (NEIS)
				PV A 2.2s 76.4nm
				LmH B 19 2.2 μm
				LmV B 19 1.7 μm
12.	eP	A	11 35 07	Hokkaido, Japan Region 41.47 N 142.00 E
				H = 11 23 12.5 h = 71 km MB = 4.8
				D = 78.53 Az = 330.3 (NEIS)
12.	eP	A	14 23 14.5	Tadzhik-Sinkiang Border Region
	LmH	B	42.0	39.39 N 73.86 E
	LmV	B	43.0	H = 14 14 54.1 h = normal MB=5.1 MS=4.7
				D = 44.18 Az = 306.1 (NEIS)
				PV A 1.6s 22.0nm M = 4.7
				LmH B 18 1.6 μm 5.0
				LmV B 13 0.8 μm 4.9
12.	eP	A	14 26 45	North of Ascension Island 0.27 S 16.53 W
				H = 14 17 04.0 h = 27 km MB = 4.9
				D = 56.10 Az = 21 (ISC)
12.	eP	A	21 25 58.5	Southern Sinkiang Prov., China
	LmH	B	45.1	39.22 N 74.04 E
	LmV	B	46.3	H = 21 17 47.6 h ≐ 27 km MB=5.2 MS=5.4
				D = 44.40 Az = 306.2 (NEIS)
				LmH B 17s 2.6 μm M = 5.2
				LmV B 16 1.3 μm 5.0

Day	Phase		h m s	Remarks
12.	eP	A	22 05 25.5	Tadzhik-Sinkiang Border Region 39.36 N 73.93 E H = 21 57 17.6 h = normal MB = 5.0 D = 44.24 Az = 306.1 (NEIS)
13.	eP	AB	03 58 13	Andreanof Islands, Aleutian Is.
	eS	B	04 08 06	51.53 N 178.11 W
	LmH	B	37.1	H = 03 46 20.3 h = 52 km MB = 5.8
	LmV	B	40.6	D = 77.87 Az = 353.7 (NEIS) PV A 1.8s 162.0nm M = 5.7 PV B 10 1.8/um 6.0 LmH B 18 6.3/um 6.0 LmV B 17.5 6.3/um 6.0
13.	ePKP	A	06 12 41.5	Fiji Islands Region 16.06 S 179.38 W H = 05 53 07.4 h = 29 km MB=5.3 MS=5.4 D = 144.38 Az = 348.0 (NEIS) PKPV A 2.0s 29.9nm
13.	ePKP	A	07 39 53.5	Fiji Islands Region 15.52 S 179.11 W H = 07 20 17.0 h = normal MB=4.9 MS=5.2 D = 143.91 Az = 348.4 (NEIS) PV A 2.2s 76.4nm
13.	e(P)	A	12 59 04	
13.	ePKP	A	13 12 21	Fiji Islands Region 15.82 S 179.48 W
	LmV	B	14 28.8	H = 12 52 47.3 h = 55 km MB = 5.4
	LmH	B	29.6	D = 144.13 Az = 347.9 (NEIS) LmH B 17s 1.7/um LmV B 17 1.7/um
13.	e	A	14 01 15	Fiji Region 15.95 S 179.29 W H = 13 41 29 h = 31 km MB = 4.9 D = 144.29 Az = 348 (ISC)
13.	eP	A	15 04 31.5	

Day	Phase		h m s	Remarks
13.	ePKIKP	A	15 22 02	New Britain Region 5.33 S 150.79 E H = 15 03 14.8 h = 100 km MB = 5.5 D = 123.43 Az = 330.1 (NEIS)
13.	eP	A	21 27 28	Tadzhik-Sinkiang Border Region 39.35 N 73.89 E H = 21 19 17.5 h = normal MB = 4.7 D = 44.23 Az = 306.1 (NEIS) traces
14.	eP	A	01 39 12	Tadzhik-Sinkiang Border Region 39.39 N 73.76 E H = 01 31 02.5 h = normal MB = 4.5 D = 44.12 Az = 306.0 (NEIS)
14.	ePKP2	A	03 33 21	Kermadec Islands Region 31.6 S 178.0 W H = 03 12 35 h = 0 km D = 159.70 Az = 342 (ISC) traces
14.	+eP	AB	05 46 46.5	Andreanof Islands, Aleutian Is. 51.56 N 178.15 W
	LmH	B	06 25.5	H = 05 34 54.4 h = 56 km MB = 5.7
	LmV	B	25.5	D = 77.83 Az = 353.7 (NEIS) PV A 1.5s 105.0nm M = 5.6 LmH B 19 1.37/um LmV B 20 1.2/um
14.	+eP	A	07 01 24	Hokkaido, Japan Region 41.13 N 142.78 E
	LmH	B	39.5	H = 06 49 21.7 h = 41 km MB = 5.3 D = 79.13 Az = 330.8 (NEIS) PV A 1.4s 32.6nm M = 5.1
14.	eP	A	15 06 54	Western Siberia 68.91 N 75.90 E
	e	A	07 10	H = 14 59 58.3 h = 0 km MB = 5.5 D = 35.03 Az = 272.2 (NEIS) PV A 1.1s 48.5nm M = 5.2 Probably underground explosion M=5.8 (UPP)

Day	Phase		h m s	Remarks
14.	eP	A	16 09 20	Crete 35.46 N 23.07 E
				H = 16 05 19.5 h = 60 km MB = 4.2
				D = 17.30 Az = 334.8 (NEIS)
14.	ePKP	A	21 34 13	Tonga Islands 15.92 S 173.13 W
	e	A	34 22	H = 21 14 37.7 h = normal MB = 4.5 MS=4.3
	e	A	34 24	D = 145.16 Az = 354.7 (NEIS)
	LmH	B	22 34.0	PV A 1.5s 22.5nm
	LmV	B	34.8	LmH B 18 1.4 µm M = 5.6
				LmV B 15 0.6 µm 5.5
14.	eP	A	22 15 02	Tadzhik-Sinkiang Border Region
				39.20 N 73.87 E
				H = 22 06 52.9 h = normal MB = 5.0
				D = 44.30 Az = 306.2 (NEIS)
15.	ePKHKP	A	01 39 00	Tonga Islands Region 22.09 S 175.68 W
	ePKP2	A	39 11	H = 01 19 24.8 h = 170 km MB = 4.8
				D = 150.93 Az = 350.4 (NEIS)
				PKHKPV A 1.7s 21.2nm
15.	ePKIKP	A	03 55 03.5	Solomon Islands 9.05 S 159.13 E
	e	A	55 36	H = 03 35 58.0 h = 59 km MB = 5.5
	ePP	A	57 18	D = 130.62 Az = 333.2 (NEIS)
	LmH	B	04 57.4	PPV A 1.8s 40.5nm M = 5.2
	LmV	B	57.5	
15.	ePKHKP	A	08 47 16	Fiji Region 20.09 S 177.5 W
				H = 08 28 31.2 h = 554 km MB = 4.4
				D = 148.65 Az = 349 (ISC)
15.	ePn	A	17 17 53	Switzerland 46.5 N 7.3 E
	e	A	18 00	H = 17 16 20 (BCIS)
	ePg	A	18 15	D = 5.02
	eSg	A	19 00	

Day	Phase		h m s	Remarks
16.	eP	A	00 19 14.5	Tadzhik-Sinkiang Border Region
				39.28 N 73.84 E
				H = 00 11 08.0 h = 50 km MB = 4.9
				D = 44.23 Az = 306.1 (NEIS)
				PV A 1.0s 15.7nm M = 4.7
16.	eP	A	01 59 29	Kurile Islands 46.78 N 153.10 E
				H = 01 47 35.1 h = normal MB = 4.4
				D = 77.42 Az = 336.0 (NEIS)
16.	ePKHKP	A	07 23 00.5	Fiji Islands Region 20.44 S 178.38 W
	ePKP2	A	23 06.5	H = 07 04 09.6 h = 503 km MB = 4.6
				D = 148.84 Az = 347.7 (NEIS)
				PKHKPV A 1.2s 24.4nm
16.	+eP	AB	09 53 26	Andreanof Islands, Aleutian Is.
	ePP	B	56 20	51.50 N 177.83 W
	ePPP	B	58 12	H = 09 41 31.7 h = 46 km MB=5.7 MS=5.8
	eS	B	10 03 16	D = 77.92 Az = 353.9 (NEIS)
	LmH	B	31.0	PV A 1.6s 99.0nm M = 5.6
	LmV	B	37.7	PV B 9 1.1 μm 5.8
				LmH B 20 4.6 μm 5.8
				LmV B 18 3.3 μm 5.7
16.	e(Sg)	A	12 36 35	
16.	e(PKP2)	A	17 16 37	Fiji Islands Region 19.81 S 177.82 W
				H = 16 57 36.4 h = 435 km MB = 4.7
				D = 148.33 Az = 348.5 (NEIS)
16.	eP	A	19 57 27	Philippine Islands Region
				6.17 N 126.96 E
				H = 19 43 46.1 h = 97 km MB = 5.4
				D = 100.87 Az = 324.1 (NEIS)
17.	+eP1	AB	05 24 07.5	Sakhalin Island 54.93 N 143.95 E
	eP2	AB	24 12	H = 05 13 08.1 h = 1 km MB=5.4 MS=4.9
	LmH	B	57.5	D = 67 53 Az = 329.4 (NEIS)

Day	Phase		h m s	Remarks
cont. 17.	LmV	B	05 57.7	P1V A 1.3s 43.7nm M = 5.5
				P2V A 1.5 70.4nm 5.7
				LmH B 15 0.6 μm 5.0
				LmV B 14 0.8 μm 5.1
17.	LmH	C	17 23.0	<u>Molucca Passage</u> 1.04 N 126.10 E
	LmV	C	32.5	H = 16 18 23.1 h = 70 km MB = 5.2 (ISC)
				D = 104.5
				LmV C 24s 0.3 μm
17.	eP	A	23 59 07.5	<u>Tadzhik-Sinkiang Border Region</u>
	LmH	B	24 18.0	39.23 N 73.91 E
	LmV	B	18.8	H = 23 50 58.9 h = 32 km MB=5.0 MS=5.3
				D = 44.30 Az = 306.2 (NEIS)
				PV A 1.6s 13.7nm M = 4.5
				LmH B 18 3.4 μm 5.3
				LmV B 18 1.4 μm 5.0
18.	eP diff	B	10 59 06	<u>Near Coast of Central Chile</u>
	ePKIKP	A	11 02 54.5	38.45 S 73.43 W
	ePP	AB	03 50	H = 10 44 12.8 h = 36 km MB=5.9 MS=7.1
	e	A	06 18	D = 115.75 Az = 44.8 (NEIS)
	eS diff	B	11 50	PdiffV B 15s 1.8 μm
	iPS	B	13 44	PKIKPV A 2.0 55.6nm
	LmH	B	49.0	PPV A 4.5 1410.0nm M = 7.0
	LmV	B	55.0	LmH B 21 80.0 μm 7.3
				LmV B 18 87.0 μm 7.4
18.	eP	A	16 11 27	<u>Tadzhik-Sinkiang Border Region</u>
				39.41 N 73.79 E
				H = 16 03 18.3 h = 21 km MB = 4.7
				D = 44.12 Az = 306.0 (NEIS)
18.	ePcP	A	17 28 36	<u>Rat Islands, Aleutian Is.</u>
				50.55 N 175.10 E
				H = 17 16 26.0 h = normal MB=5.0 MS=4.7
				D = 78.20 Az = 349.3 (NEIS)

Day	Phase		h m s	Remarks
18.	LmH	B	21 02.0	Off Coast of Southern Chile
	LmV	B	02.0	45.71 S 76.5 W
				H = 19 53 36 h = 107 km MB = 4.6 (ISC)
				D = 122.3
				LmH B 20s 0.4/um
				LmV B 19 0.55/um
18.	LmH	C	24 16.0	Off Coast of Southern Chile
	LmV	C	21.0	41.62 S 76.32 W
				H = 23 07 49.0 h = 27 km MB = 5.2 (ISC)
				D = 119.4
				LmH C 19s 0.25/um M = 4.9
				LmV C 18 0.35/um 5.0
19.	eP	A	12 30 05.5	South of Honshu, Japan
	LmH	B	13 07.3	33.31 N 139.49 E
	LmV	B	12.2	H = 12 17 32.5 h = 23 km MB=5.3 MS=4.6
				D = 84.60 Az = 329.7 (NEIS)
				PV A 1.4s 39.6nm M = 5.5
				LmH B 14 0.9/um 5.3
				LmV B 12 0.5/um 5.2
19.	eP	A	20 07 25	Off Coast of Central America
	epP	A	07 36	12.31 N 88.93 W
	e(PP)	A	11 02	H = 19 54 44.8 h = 67 km MB = 5.2
	LmH	B	52.0	D = 87.17 Az = 38.8 (NEIS)
	LmV	B	52.0	h = 40 km
				PV A 1.4s 20.9nm M = 5.1
				PPV A 2.0 34.2nm 5.5
				LmH B 16 0.5/um
				LmV B 16 0.35/um
20.	eP	A	04 52 23	South of Honshu, Japan
	LmH	B	C¯ 35.0	33.42 N 139.41 E
				H = 04 39 52.8 h = 31 km MB = 4.8
				D = 84.47 Az = 329.6 (NEIS)

Day	Phase		h m s	Remarks
20.	LmH	B	07 11.0	East China Sea 27.10 N 127.0 E
	LmV	B	18.0	H = 06 23 02 h = 52 km MB = 4.4 (ISC)
				D = 83.8
				LmH B 18s 1.8 μm
				LmV B 15 0.9 μm
20.	LmH	B	15 41.5	Tadzhikistan-Sinkiang Border Region
				39.32 N 73.64 E
				H = 15 14 36.6 h = 43 km MB = 4.3 (ISC)
				D = 44.2
				LmH B 18s 0.3 μm M = 4.3
20.	LmH	B	19 38.0	Western Caroline Islands
	LmV	B	38.2	11.26 N 140.81 E
				H = 18 31 20.4 h = 39 km MB = 5.2 (ISC)
				D = 104.2
				LmH B 17s 0.6 μm M = 5.2
				LmV B 17 0.7 μm 5.3
20.	+iP	AB	20 56 46.5	Rat Islands, Aleutian Is.
	ePP	A	59 43.5	52.24 N 174.97 E
	ePS	B	21 07 28	H = 20 45 01.4 h = 58 km MB = 5.6
	LmH	B	28.5	D = 76.52 Az = 349.2 (NEIS)
	LmV	B	28.5	PV A 1.2s 69.2nm M = 5.5
				PPV A 1.7 48.5nm 5.5
				LmH B 23 1.2 μm
				LmV B 23 1.2 μm
20.	eP	A	23 56 00	Greece 38.24 N 20.78 E
	e	A	56 21.5	H = 23 52 40.1 h = 50 km MB = 4.5
				D = 14.01 Az = 335.2 (NEIS)
21.	LmH	B	03 10.5	LmH B 17s 0.8 μm
	LmV	B	17.7	LmV B 15 0.5 μm
21.	eP	A	13 06 11.5	Ionian Sea 37.24 N 19.61 E
	LmH	C	12.9	H = 13 02 46.2 h = normal MB = 4.1

Day	Phase		h m s	Remarks
cont. 21.	LmV	C	13 12.9	D = 14.57 Az = 339.4 (NEIS)
				LmH C 11s 0.6 µm M = 4.0
				LmV C 12 0.4 µm 4.0
21.	eP	A	15 10 25	Eastern Mediterranean Sea
				33.8 N 25.41 E
				H = 15 05 52 h = 0 km
				D = 19.60 Az = 333 (ISC)
21.	eP	A	18 53 25	Southern Sinkiang Prov., China
	LmH	C	19 13.7	39.23 N 74.03 E
	LmV	C	13.7	H = 18 45 16.7 h = normal MB = 5.0
				D = 44.38 Az = 306.2 (NEIS)
				PV A 1.1s 18.2nm M = 4.8
				LmH C 10 0.45 µm 4.7
				LmV C 10 0.45 µm 4.8
21.	eP	A	21 26 33	Molucca Passage 0.42 N 125.15 E
	e	A	26 58	H = 21 12 29.9 h = 46 km MB=5.0 MS=4.6
	LmH	C	22 12.0	D = 104.39 Az = 322.9 (NEIS)
	LmV	C	13.0	PV A traces
21.	LmH	C	24 53.0	Off Coast of Northern Chile
	LmV	C	53.0	19.86 S 71.17 W
				H = 23 59 00.7 h = 44 km MB=5.1 (ISC)
				D = 100.7
				LmH C 22s 0.25 µm M = 4.7
				LmV C 20 0.3 µm 4.8
22.	ePKIKP	A	12 07 54	Fiji Islands Region 20.72 S 178.47 W
	ePKHKP	A	07 59.5	H = 11 49 14.8 h = 583 km MB = 5.1
	ePKP2	A	08 06	D = 149.09 Az = 347.5 (NEIS)
				PKHKPV A 1.3s 39.3nm
23.	eP	A	04 11 29.5	Taiwan 23.83 N 121.56 E
	LmH	B	53.5	H = 03 58 49.1 h = normal MB = 5.0

Day	Phase		h m s	Remarks
cont. 23.	LmV	B	04 53.5	D = 83.63 Az = 323.0 (NEIS)
				PV A 1.6s 19.3nm M = 5.0
				LmH B 15 0.6 um 5.1
				LmV B 15 1.0 um 5.4
23.	+ePKIKP	A	05 08 58	<u>Banda Sea</u> 7.49 S 127.51 E
	ePP1	A	09 44	H = 04 50 34.6 h = 136 km MB = 5.8
	ePP2	A	09 48	D = 112.06 Az = 321.8 (NEIS)
	ePS	C	18 55	PKIKPV A 1.2s 32.5nm
	ePKKP	A	19 50	PP2V A 2.0 94.0nm M = 5.8
	ePPS	B	20 10	LmH B 19 1.5 um
	e	B	21 16	LmV B 19 1.0 um
	eSS	C	25 25	
	LmH	B	54.5	
	LmV	B	57.5	
23.	ePKP	A	06 06 45	<u>Tonga Islands</u> 16.99 S 173.48 W
				H = 05 47 06.9 h = normal MB = 4.4
				D = 146.18 Az = 354.2 (NEIS)
				PKPV A 1.3s 15.3nm
23.	LmH	C	16 55.0	<u>Tadzhikistan-Sinkiang Border Region</u>
	LmV	C	55.0	39.34 N 73.68 E
				H = 16 26 32.8 h = 50 km MB = 4.7 (ISC)
				D = 44.2
23.	LmH	C	17 52.0	<u>Northern Easter I. Cordillera</u>
	LmV	C	55.0	4.58 S 105.54 W
				H = 16 52 03.4 h = 33 km MB = 4.8 (ISC)
				D = 110.5
23.	LmH	C	24 30.0	<u>North Atlantic Ocean</u> 19.22 N 68.04 W
	LmV	C	30.3	H = 23 55 35 h = 15 km MB = 4.9 (ISC)
				D = 68.8
24.	+eP	AB	03 00 01	<u>Colombia</u> 4.28 N 76.86 W
				H = 02 47 30.1 h = 84 km MB = 5.9
				D = 85.75 Az = 39.7 (NEIS)
				PV A 1.4s 97.7nm M = 5.6

Day	Phase		h m s	Remarks
24.	ePKHKP	A	03 41 33.5	**Tonga Islands** 21.71 S 174.17 W
				H = 03 21 43.8 h = normal MB=4.8 MS=4.3
				D = 150.76 Az = 352.4 (NEIS)
				PKHKPV A 1.9s 45.5nm
24.	eP	A	10 23 55	**Southern Greece** 37.75 N 21.17 E
				H = 10 20 20.5 h = normal MB = 4.2
				D = 14.58 Az = 335.2 (NEIS)
				PV A 1.3s 10.9nm M = 4.2
24.	+eiP	AB	10 53 03	**Fox Islands, Aleutian Is.**
	epP	A	53 15	52.41 N 168.27 W
	eS	B	11 03 00	H = 10 41 11.2 h = 41 km MB=5.7 MS=5.6
	eSS	C	08 00	D = 77.32 Az = 0.1 (NEIS)
	LmH	B	32.3	h = 41 km
	LmV	B	32.3	PV A 1.4s 149.0nm M = 5.8
				PV B 5 1.3μm 6.2
				LmH B 19 1.7μm 5.4
				LmV B 19 1.9μm 5.5
24.	eP	A	11 29 53.5	**Tadzhik-Siniang Border Region**
	ePP	A	31 42.5	39.34 N 73.67 E
				H = 11 21 46.8 h = normal MB = 4.9
				D = 44.09 Az = 306.1 (NEIS)
				PV A 1.2s 14.2nm M = 4.7
24.	eP	A	12 22 47	**Tadzhik-Sinkiang Border Region**
				39.35 N 73.88 E
				H = 12 14 37.1 h = normal MB = 4.9
				D = 44.21 Az = 306.1 (NEIS)
				traces
24.	eP	A	18 29 38	**Burma** 25.81 N 96.24 E
				H = 18 18 45.8 h = normal MB = 4.7 (NEIS)
				D = 67.19

Day	Phase		h m s	Remarks
24.	ePKHKP	A	18 52 09.5	Tonga Islands 21.52 S 174.51 W
	ePKP2	A	52 20.5	H = 18 32 19.3 h = normal MB = 4.8
				D = 150.54 Az = 352.1 (NEIS)
				PKHKPV A 1.7s 36.3nm
24.	ePKHKP	A	18 52 09.5	Tonga Islands 21.52 S 174.51 W
	ePKP2	A	51 20.5	H = 18 32 19.3 h = normal MB = 4.8
				D = 150.54 Az = 352.1 (NEIS)
				PKHKPV A 1.7s 36.3nm
24.	eP	A	22 12 09	Ionian Sea 37.89 N 19.64 E
	LmH	B	18.7	H = 22 08 38.6 h = 11 km MB = 4.1
	LmV	B	18.7	D = 13.97 Az = 338.4 (NEIS)
				LmH B 10.5s 0.45/um M = 3.8
				LmV B 11 0.65/um
24.	+iP	AB	22 30 48	Fox Islands, Aleutian Is.
	LmH	C	23 07.0	52.30 N 168.31 W
	LmV	C	15.4	H = 22 18 55.4 h = 37 km MB=5.3 MS=4.5
				D = 77.43 Az = 0.0 (NEIS)
				PV A 1.3s 63.3nm M = 5.5
25.	-iP	AB	01 31 22.3	South of Honshu, Japan
	ePP	B	34 44	32.05 N 142.29 E
	ePP	A	34 45	H = 01 18 39.9 h = normal MB=5.9 MS=5.6
	e(S)	BC	41 58	D = 86.86 Az = 331.1 (NEIS)
	eSS	C	48 15	PV A 1.5s 181.0nm M = 6.1
	LmH	B	02 07.4	PV B 6 1.6/um 6.5
	LmV	B	18.6	PPV A 2.4 172.7nm 6.1
				PPV B 7 1.2/um 6.4
				LmH B 18 5.6/um 6.0
				LmV B 17 5.3/um 6.0
25.	eP	A	02 24 36	South of Honshu, Japan
				32.07 N 142.41 E
				H = 02 11 51.7 h = normal MB = 4.9
				D = 86.89 Az = 331.1 (NEIS)
				traces

Day	Phase		h m s	Remarks
25.	ePKIKP	A	03 47 23	Fiji Islands Region 16.84 S. 175.78 E H = 03 27 46.3 h = normal MB=5.1, MS=5.3 D = 143.96 Az = 342.8 (NEIS) PKIKPV A 1.3s 8.7nm
25.	eP epP LmV	A A B	04 32 00 32 11.5 05 01.0	South of Honshu, Japan 31.96 N 142.41 E H = 04 20 17.0 h = 40 km MB = 5.3 (NEIS) D = 86.98 h = 45 km PV A 1.2s 22.4nm M = 5.3 LmV B 17 0.7/um 5.2
25.	eP	A	06 00 02	Ionian Sea 37.72 N 20.08 E H = 05 56 36.9 h = 9 km MB = 3.9 D = 14.26 Az = 337.7 (NEIS) traces
25.	eP e	A A	10 07 50 08 02	South of Honshu, Japan 31.91 N 142.47 E H = 09 55 11.7 h = 81 km MB = 5.2 D = 87.05 Az = 331.2 (NEIS) PV A 1.5s 20.1nm M = 5.0
25.	eP LmH LmV	A C C	10 26 07.5 55.0 55.0	South of Honshu, Japan 32.04 N 142.28 E H = 10 13 18.1 h = normal MB = 5.0 D = 86.86 Az = 331.1 (NEIS)
25.	ePKHKP	A	12 10 52	Fiji Islands Region 19.75 S 177.99 W H = 11 51 51.2 h = 404 km MB = 4.8 D = 148.24 Az = 348.4 (NEIS)
25.	ePP	A	13 35 08.5	Greece 38.34 N 19.97 E H = 13 31 23.4 h = normal MB = 4.0 D = 13.66 Az = 336.9 (NEIS)
25.	ePKIKP ePKHKP	AB AB	14 53 33 53 38	South of Fiji Islands 23.48 S 179.91 W H = 14 34 46.7 h = 542 km MB = 5.3

Day	Phase		h m s	Remarks
cont. 25.	ePKP2	A	14 53 51	D = 151.44 Az = 344.6 (NEIS)
	epPKIKP	A	55 44	h = 575 km
	epPKHKP	AB	55 47.5	PKIKPV A 1.8s 54.0nm
	epPKP2	A	56 01	PKHKPV A 1.6 110.0nm
	e	C	15 03 23	PKP2V A 1.3 30.6nm
				pPKIKPV A 1.8 121.6nm
				pPKHKPV A 1.5 183.4nm
				pPKP2V A 1.5 55.3nm
25.	eP	A	19 46 28	<u>Ionian Sea</u> 38.43 N 19.78 E
	e	A	46 35	H = 19 43 11.5 h = 33 km MB = 4.1 (NEIS)
	LmH	C	53.0	D = 13.52
	LmV	C	53.0	
26.	+iP	A	06 40 18.4	<u>Mexico-Guatemala Border Region</u>
	LmH	C	07 20.0	15.95 N 91.04 W
	LmV	C	20.0	H = 06 27 40.7 h = normal MB=5.1 MS=4.8
				D = 85.62 Az = 38.5 (NEIS)
				PV A 1.0s 27.6nm M = 5.4
26.	eP	A	18 13 57	<u>Mediterranean Sea</u> 35.25 N 22.27 E
	eX	A	14 22	H = 18 09 55.9 h = 33 km MB = 3.8
				D = 17.22 Az = 336.6 (NEIS)
				XV A 1.5s 12.6nm
27.	eP	A	05 51 42.5	<u>Southern Sinkiang Prov., China</u>
	e	A	52 17	39.28 N 73.96 E
				H = 05 43 33.7 h = normal MB = 5.0
				D = 44.31 Az = 306.2 (NEIS)
27.	ePKIKP	A	06 42 48	<u>Near Coast of Central Chile</u>
	LmH	B	07 32.0	38.34 S 73.39 W
	LmV	B	32.0	H = 06 24 07.1 h = 23 km MB=5.4 MS=4.6
				D = 115.64 Az = 44.7 (NEIS)
				LmH B 19s 0.7 μm M = 5.3
				LmV B 20 0.6 μm 5.2

Day	Phase		h m s	Remarks
27.	eP	AB	13 04 09	Tadzhik-Sinkiang Border Region
	Pm	A	04 14	39.67 N 73.83 E
	ePP	B	05 59	H = 12 56 03.2 h = normal MB=5.8 MS=5.9
	eS	B	10 45	D = 44.00 Az = 305.8 (NEIS)
	eSS	B	13 46	PmV A 1.3s 216.0nm M = 5.8
	LmH	B	23.2	PPV B 8.5 1.3 µm 5.8
	LmV	B	24.4	LmH B 18 23.9 µm 6.2
				LmV B 14 27.1 µm 6.4
27.	eP	A	14 32 52.5	Southern Sinkiang Prov., China
				39.62 N 74.03 E
				H = 14 24 45.4 h = normal MB = 4.2
				D = 44.15 Az = 305.9 (NEIS)
27.	ePP	A	15 38 53	Catamarca Province, Argentina
				27.93 S 66.67 W
				H = 15 20 49.8 h = 147 km MB = 5.5 (NEIS)
				D = 104.18
27.	eP	AB	17 42 05.5	Tadzhik-Sinkiang Border Region
	ePP	A	43 53	39.37 N 73.86 E
	e	A	44 07	H = 17 33 58.1 h = normal MB=5.3 MS=5.2
	eSS	C	52 00	D = 44.19 Az = 306.1 (NEIS)
	LmH	B	18 01.0	PV A 1.7s 48.5nm M = 5.0
	LmV	B	02.4	LmH B 19 2.8 µm 5.2
				LmV B 15 1.9 µm 5.2
28.	eP	A	01 02 23.5	Near East Coast of Honshu, Japan
	epP	A	02 35	36.48 N 141.10 E
				H = 00 50 04.3 h = 51 km MB = 5.0
				D = 82.52 Az = 330.3 (NEIS)
				h = 44 km
				PV A 1.0s 13.8nm M = 4.9
28.	ePKP	A	10 37 59	Fiji Islands Region 17.89 S 178.51 W
				H = 10 19 27.4 h = 613 km MB = 4.8
				D = 146.33 Az = 348.4 (NEIS)
				PKPV A 1.6s 24.8nm

Day	Phase		h m s	Remarks
28.	eP	A	18 54 32	Gulf of Alaska 59.51 N 144.45 W
	eS	B	19 03 35	H = 18 43 25.7 h = 4 km MB=4.9 MS=4.6
	LmH	B	30.5	D = 68.50 Az = 16.1 (NEIS)
	LmV	B	31.4	PV A 1.3s 35.0nm M = 5.4
				LmH B 15 0.7 µm 5.0
				LmV B 14 1.0 µm 5.3
29.	-iP	A	01 10 17.4	Afghanistan-USSR Border Region
				36.52 N 71.33 E
				H = 01 02 28.5 h = 228 km MB = 5.0
				D = 44.31 Az = 308.1 (NEIS)
				PV A 1.2s 36.6nm M = 4.7
29.	ePKP2	A	03 10 34	Kermadec Islands Region
	e	A	10 50	28.92 S 177.47 W
	e	A	10 58	H = 02 50 14.9 h = 58 km MB = 5.3
	e	A	11 10	D = 157.24 Az = 344.9 (NEIS)
29.	ePKHKP	A	04 52 18	Tonga Islands 19.24 S 173.35 W
	ePKP2	A	52 21.5	H = 04 32 33.3 h = normal MB = 5.2 MS=4.8
				D = 148.42 Az = 354.0 (NEIS)
				PKHKPV A 1.9s 60.6nm
29.	eP	A	06 52 04.5	Rat Islands, Aleutian Is.
	e	A	52 22	50.24 N 177.65 E
				H = 06 40 03.3 h = normal MB = 4.9
				D = 78.79 Az = 351.0 (NEIS)
29.	+iP	AB	10 06 01.7	Novaja Zemlya 73.37 N 55.09 E
	ePP	B	07 15	H = 09 59 55.6 h = 0 km MB=6.4 MS=5.0
	eS	AB	11 33	D = 29.36 Az = 243.3 (NEIS)
	LmH	B	21.7	Underground explosion MB = 6.5 (UPP)
	LmV	B	21.8	PV A 1.4s 655.0nm M = 6.3
	eP'P'	A	39 17	PH A 1.4 515.0nm 6.3
				PV B 4 1.7 µm 6.3
				PPV B 3.5 0.9 µm 6.2
				LmH B 7 12.7 µm 6.0
				LmV B 7 9.6 µm 6.0

Day	Phase		h m s	Remarks
29.	eP	A	15 06 10	Ural Mountains Region 67.23 N 62.12 E
	e	A	06 35.5	H = 14 59 59.6 h = 0 km MB = 5.2
				D = 29.78 Az = 261.6 (NEIS)
				Underground explosion MB = 5.4 (UPP)
30.	epP	A	07 58 17	Near Coast of Nicaragua
				12.76 N 87.38 W
				H = 07 45 20.9 h = 62 km MB = 4.9
				D = 85.87 Az = 39.1 (NEIS)
30.	+eP	A	15 12 18	Southern Nevada 37.15 N 116.08 W
	ePP	A	16 22	H = 15 00 00.2 h = 0 km MB = 5.8
				D = 81.22 Az = 30.6 (NEIS)
				Nuclear explosion (AEC)
				PV A 1.4s 65.0nm M = 5.5
30.	ePn	A	17 45 08.5	Yugoslavia 44.17 N 20.37 E
	e(Sg)	A	47 30	H = 17 42 59.4 h = normal
	e	A	47 51	D = 8.78 Az = 320.6 (NEIS)
30.	eP1	AB	23 42 13	South of Honshu, Japan
	eP2	A	42 16	30.56 N 141.94 E
	ePP	A	45 40	H = 23 29 23.6 h = 24 km MB=5.3 MS=5.7
	eX	A	46 03	D = 88.01 Az = 330.9 (NEIS)
	eSKS	B	52 40	P1V A 1.4s 13.9nm M = 5.1
	eS	B	52 56	P1V B 6 0.7/um 6.2
	LmH	B	24 19.4	P2V A 2.5 131.0nm 5.8
	LmV	B	25.1	XV A 2.8 191.5nm
				LmH B 19 3.9/um 5.9
				LmV B 18 2.9/um 5.8
31.	+ePKP	A	01 33 44.5	Loyalty Islands Region
				22.66 S 170.93 E
				H = 01 14 00.5 h = 17 km MB = 4.6
				D = 147.74 Az = 335.1 (NEIS)
				PKPV A 1.0s 23.6nm

Day	Phase		h m s	Remarks
31.	LmH	C	17 42.4	<u>Tadzhikistan-Sinkiang Border Region</u>
	LmV	C	42.4	39.54 N 73.87 E
				H = 17 13 58.5 h = 51 km MB = 4.2 (ISC)
				D = 44.3
31.	LmH	C	19 05.5	<u>Northern Sumatra</u> 0.62 N 97.95 E
	LmV	C	05.5	H = 18 05 08 h = 63 km MB = 4.5 (ISC)
				D = 87.3
31.	e	A	23 53 25	

Day	Phase		h m s	Remarks
1.	LmH	C	10 40.0	West Chile Rise 42.85 S 82.5 W H = 09 33 42.4 h = 33 km MB = 5.0 (ISC) D = 123.8 LmH C 24s 0.4/um M = 5.0
1.	eP	A	11 05 51	Jan Mayen Island Region 71.85 N 2.27 W H = 11 00 58.1 h = normal MB = 4.1 D = 22.18 Az = 156.1 (NEIS) PV A 1.8s 30.4nm M = 4.4
1.	eP	A	13 20 05	Jan Mayen Island Region 71.86 N 2.25 W H = 13 15 11.7 h = normal MB = 3.9 D = 22.18 Az = 156.2 (NEIS) PV A 1.6s 24.7nm M = 4.4
1.	eSn eSg	A A	17 26 15 27 20	France 45.8 N 1.5 W H = 17 21 57 (BCIS) D = 10.04
2.	-eP LmH	A C	04 46 21 05 33.5	Southern Sumatra 2.77 S 101.21 E H = 04 33 16.6 h = 52 km MB = 5.7 D = 91.87 Az = 320.4 (NEIS) PV A 1.4s 83.8nm M = 6.0 LmH C 21 0.3/um
2.	eP ePP	A A	08 54 59 58 38	Southern Sumatra 2.77 S 101.24 E H = 08 41 54.9 h = 56 km MB = 5.4 D = 91.89 Az = 320.4 (NEIS) PV A 1.6s 44.0nm M = 5.6
2.	e	A	23 29 54	Greece 38.5 N 23.7 E H = 23 26 08 (BCIS) D = 14.92 traces

Day	Phase		h m s	Remarks
3.	eSg	A	00 20 21.5	<u>Switzerland</u> 47°27'N 7°38'E
				H = 00 18 10 (BCIS)
				D = 4.12
3.	eP	A	01 52 30	<u>South of Honshu, Japan</u>
	e	A	53 02	32.16 N 142.30 E
	e	A	57 25	H = 01 39 45.2 h = 20 km MB=5.3 MS=4.9
	LmH	B	02 37.2	D = 86.76 Az = 331.1 (NEIS)
	LmV	B	37.3	PV A 2.0s 34.2nm M = 5.2
				LmH B 16 1.7µm 5.5
				LmV B 14.5 1.2µm 5.5
3.	eP	A	01 57 41	<u>Kurile Islands</u> 49.67 N 156.01 E
				H = 01 45 56.9 h = 22 km MB=5.2 MS=4.9
				D = 75.53 Az = 337.5 (NEIS)
				PV A 1.1s 16.1nm M = 5.0
3.	eSg	A	05 55 16	<u>Switzerland</u> 47.44 N 7.62 E
				H = 05 53 03.2 h = 0 km
				D = 4.15 Az = 38 (ISC)
3.	eiP	AB	06 07 52	<u>Philippine Islands Region</u>
	+eiX	A	08 04.5	18.34 N 119.17 E
	ePP	AB	11 15	H = 05 55 06.4 h = 11 km MB=5.9 MS=5.4
	eS	C	18 20	D = 86.63 Az = 322.6 (NEIS)
	ePS	B	19 42	PV A 1.4s 176.7nm M = 6.1
	LmH	B	46.6	XV A 1.4 93.0nm
	LmV	B	53.3	LmH B 15.5 3.7µm 5.9
				LmV B 14.4 3.5µm 5.9
3.	eP	A	06 12 23	<u>Philippine Islands Region</u>
				18.21 N 119.13 E
				H = 05 59 41.0 h = normal MB = 5.6
				D = 86.71 Az = 322.5 (NEIS)
				PV A 1.5s 35.2nm M = 5.4
3.	+eP	A	19 49 27	<u>Tadzhik-Sinkiang Border Region</u>
	ePP1	A	51 04	39.43 N 73.70 E

Day	Phase		h m s	Remarks
cont. 3.	ePP2	A	19 51 16	H = 19 41 19.9 h = normal MB = 5.4
	LmH	B	20 08.7	D = 44.06 Az = 306.0 (NEIS)
	LmV	B	09.7	PV A 1.8s 81.1nm M = 5.2
				PP1V A 1.9 102.3nm 5.3
				PP2V A 2.0 119.7nm 5.4
				LmH B 15.5 1.5μm 5.0
				LmV B 12 2.0μm 5.3
4.	eP	A	06 18 33.5	<u>South of Java</u> 9.03 S 107.57 E
	ePP	A	22 39	H = 06 04 50.7 h = 69 km MB = 5.4
				D = 100.69 Az = 319.9 (NEIS)
				PV A 1.6s 16.5nm M = 5.4
				PPV A 1.8 27.0nm 5.6
4.	eP1	AB	06 33 22	<u>Mediterranean Sea</u> 33.13 N 13.57 E
	eP2	A	33 26.5	H = 06 29 16.4 h = 17 km MB=5.1 MS=5.6
	eS	B	36 40	D = 17.56 Az = 355.9 (NEIS)
	LmH	B	40.3	P1V A 1.8s 135.0nm M = 4.8
	LmV	B	42.8	P2V A 1.8 372.0nm 5.2
				SH B 15.5 6.7μm 6.2
				LmH B 18.5 22.4μm 5.5
				LmV B 11 10.7μm 5.5
4.	ePKHKP	A	08 18 12	<u>Tonga Islands Region</u> 22.58 S 175.23 W
				H = 07 58 22.1 h = 58 km MB = 5.1
				D = 151.48 Az = 350.9 (NEIS)
4.	eP	A	09 32 05	<u>Near East Coast of Honshu, Japan</u>
	e(pP)	A	32 26	40.20 N 141.74 E
	e(sP)	A	32 41.5	H = 09 20 02.5 h = 55 km MB = 5.3
	LmH	B	10 05.5	D = 79.54 Az = 330.3 (NEIS)
	LmV	B	12.2	PV A 1.2s 16.3nm M = 4.8
				LmH B 16.5 0.9μm
				LmV B 16 0.4μm
5.	eP	A	04 05 40	<u>Central Russia</u> 50.97 N 89.39 E
				H = 03 57 14.2 h = normal MB = 4.4
				D = 46.94 Az = 301.6 (NEIS)

Day	Phase		h m s	Remarks
5.	eP	A	06 35 36	South of Honshu, Japan 32.23 N 142.34 E H = 06 22 53.0 h = normal MB = 5.1 D = 86.72 Az = 331.1 (NEIS) PV A 1.1s 12.10nm M = 5.1
5.	eP	A	07 59 40	Northern Colombia 6.80 N 73.04 W H = 07 47 40.7 h = 160 km MB = 5.4 D = 81.40 Az = 39.9 (NEIS) PV A 1.9s 30.3nm M = 4.7
5.	eP	A	10 08 09.5	Mexico-Guatemala Border Region 16.70 N 91.69 W H = 09 55 32.9 h = normal MB = 4.6 D = 85.43 Az = 38.4 (NEIS)
5.	eP	A	11 38 42.5	Crete 35.69 N 25.05 E
	eX	A	38 50	H = 11 34 36.1 h = 53 km MB = 4.2
	LmH	B	44.9	D = 17.83 Az = 331.1 (NEIS)
	LmV	B	47.3	XV A 1.8s 47.3nm LmH B 18 1.2 /um LmV B 14 0.6 /um
5.	eP	A	14 40 56.5	Afghanistan-USSR Border Region 37.28 N 71.71 E H = 14 33 00.2 h = 137 km MB = 4.5 D = 44.09 Az = 307.5 (NEIS)
5.	eP	A	18 34 10	Ryukyu Islands 29.35 N 130.51 E
	LmH	B	19 16.5	H = 18 21 42.2 h = normal MB = 4.9
	LmV	B	16.5	D = 83.84 Az = 325.9 (NEIS) PV A 1.3s 13.1nm M = 4.9 LmH B 14 0.7 /um 5.2 LmV B 15 0.8 /um 5.3
5.	LmH	B	22 03.4	New Guinea 4.26 S 143.55 E
	LmV	B	11.3	H = 21 17 57.5 h = 61 km MB = 5.8 (ISC) D = 118.8 LmH B 19s 1.2 /um LmV B 19 0.9 /um

Day	Phase		h m s	Remarks
6.	ePKIKP	A	03 16 22	<u>Solomon Islands</u> 4.08 S 154.01 E H = 02 58 10.9 h = 420 km MB = 5.2 D = 123.89 Az = 332.1 (NEIS)
6.	eP	A	15 32 05.5	<u>Tadzhik-Sinkiang Border Region</u>
	LmH	B	51.3	39.32 N 73.79 E
	LmV	B	52.5	H = 15 23 58.3 h = normal MB = 4.9 D = 44.18 Az = 306.1 (NEIS) LmH B 16s 1.2/um M = 4.9 LmV B 16 0.6/um 4.7
6.	ePKP	A	20 55 25.5	<u>Tonga Islands</u> 15.62 S 173.33 W H = 20 36 02.1 h = 124 km MB = 4.7 D = 144.83 Az = 354.5 (NEIS) PKPV A 1.0s 15.7nm
6.	ePKIKP	A	23 45 31	<u>Solomon Islands</u> 7.10 S 155.86 E
	epPKIKP	A	45 45	H = 23 26 32.8 h = 63 km MB = 5.4 (NEIS)
	LmH	C	24 33.2	D = 127.40 h = 48 km
	LmV	C	38.5	PKIKPV A 2.0s 42.7nm pPKIKPV A 1.5 47.7nm LmH C 25 0.45/um LmV C 26 0.5/um
7.	eP	A	04 21 22	<u>Hokkaido, Japan Region</u> 43.79 N 140.25 E H = 04 09 59.7 h = 240 km MB = 4.2 D = 75.88 Az = 329.1 (NEIS)
7.	eP	A	07 35 17	<u>Venezuela</u> 4.99 N 78.30 W H = 07 22 35.7 h = 22 km MB = 4.8 (NEIS) D = 86.13
7.	eP	A	15 54 40	<u>Tadzhik - Sinkiang Border Region</u>
	LmH	B	16 13.5	39.29 N 73.90 E
	LmV	B	14.0	H = 15 46 30.9 h = normal MB = 4.9 D = 44.26 Az = 306.1 (NEIS) LmH C 19s 2.8/um M = 5.2 LmV C 19 1.2/um 4.8

Day	Phase		h m s	Remarks
7.	eP	A	19 51 41	<u>Leeward Islands</u> 15.10 N 60.63 W
	epP	A	51 58.5	H = 19 40 52.2 h = 58 km MB = 5.7
				D = 67.22 Az = 41.1 (NEIS)
				h = 67 km
7.	eP	AC	20 57 06	<u>South of Java</u> 9.82 S 108.36 E
	eX	A	21 00 27	H = 20 43 11.5 h = normal MB=6.1 MS=6.5
	iPP	AB	01 16	D = 101.79 Az = 319.8 (NEIS)
	eiSKS	C	07 43	PV A 1.7s 60.6nm M = 5.9
	eS	C	08 40	XV A 2.2 109.1nm
	ePKKP	A	13 28	PPV A 2.0 316.0nm 6.5
	e	A	13 36.5	PPV B 9.5 4.0/um 6.9
	eiSS	C	15 44	LmH B 22.5 21.9/um 7.6
	LmH	B	46.8	LmV B 19 8.5/um 6.3
	LmV	B	49.1	
8.	ePKP	A	05 32 42.5	<u>New Hebrides Islands</u> 14.91 S 167.97 E
				H = 05 13 11.6 h = 17 km MB=5.5 MS=5.3
				D = 139.55 Az = 336.8 (NEIS)
8.	-eiPKIKP	A	05 35 34	<u>New Ireland Region</u> 3.69 S 153.93 E
	-eipPKIKP	A	37 22.5	H = 05 17 27.5 h = 449 km MB = 5.7
				D = 123.51 Az = 332.2 (NEIS)
				h = 470 km
				PKIKPV A 1.3s 87.3nm
8.	ePP	A	19 13 30	<u>Aegean Sea</u> 39.86 N 24.45 E
	e	A	13 54	H = 19 10 00.2 h = normal MB = 4.3
				D = 14.05 Az = 324.4 (NEIS)
8.	LmH	C	20 19.3	LmH C 10s 1.0/um
	LmV	C	19.3	LmV C 10. 0.7/um
9.	ePKP	A	00 27 42.5	<u>Fiji Region</u> 15.6 S 176.11 W
				H = 00 08 10.2 h = 33 km MB = 4.5
				D = 144.50 Az = 352 (ISC)

Day	Phase		h m s	Remarks
9.	e(PKP)	A	04 05 20	South of Fiji 23.2 S 179.7 E
				H = 03 45 50 h = 33 km (ISC)
				D = 151.1
9.	ePn	A	12 51 22.5	Northern Italy 46.22 N 12.57 E
	ePg	A	51 40	H = 12 50 15.3 h = normal
	eSn	A	52 12	D = 4.47 Az = 352.2 (NEIS)
	eSg	A	52 36	
9.	eP	A	18 05 26.5	North Atlantic Ridge 34.6 N 36.63 W
				H = 17 58 10 h = 32 km MB = 5.0
				D = 38.24 Az = 50 (ISC)
				PV A 1.8s 20.3nm M = 4.6
10.	eP	A	00 28 58	Ionian Sea 37.83 N 19.82 E
				H = 00 25 33.1 h = normal MB = 3.9
				D = 14.09 Az = 338.1 (NEIS)
10.	ePKIKP	A	21 26 04	Kermadec Islands 30.43 S 177.81 W
	epPKIKP	A	26 11	H = 21 06 07.9 h = 27 km MB=5.5 MS=5.1
	ePKP2	A	26 39	D = 158.61 Az = 343.4 (NEIS)
	e	A	26 53	h = 29 km
	ePP	A	30 16	PKIKPV A 2.2s 65.4nm
	LmH	C	22 46.0	PKP2V A 1.6 76.9nm
	LmV	C	46.0	LmH C 20 0.35/um M = 5.1
				LmV C 20 0.35/um 5.2
11.	ePKIKP	A	01 36 53	Kermadec Islands 30.36 S 177.97 W
	ePKP2	A	37 30.5	H = 01 17 02.1 h = 43 km MB=5.3 MS=5.2
	LmH	C	02(40)	D = 158.50 Az = 343.2 (NEIS)
	LmV	C	43.0	PKIKPV A 2.0s 34.2nm
				PKP2V A 1.5 45.2nm
				LmV C 26 0.6/um M = 5.3
11.	eP	A	05 15 50.5	Albania 40.09 N 19.70 E
	LmH	C	20.0	H = 05 12 57.0 h = 51 km MB = 4.4
	LmV	C	21.3	D = 11.98 Az = 334.5 (NEIS)

Day	Phase		h m s	Remarks
11.	ePKP	A	16 37 22	Tonga Islands 15.02 S 172.97 W
				H = 16 17 50.0 h = normal MB=5.2 MS=4.7
				D = 144.28 Az = 355.0 (NEIS)
11.	ePP	A	19 35 27	Flores Islands Region 8.37 S 121.94 E
	LmV	C	20 15.0	H = 19 16 17.2 h = 33 km MB = 5.8 (NEIS)
	LmH	C	18.0	D = 109.28
				PPV A 1.3s 18.2nm M = 5.6
				LmH C 32 0.6/um 4.9
				LmV C 40 0.7/um 4.9
12.	eP	A	05 31 48.5	Off Coast of Northern California
				41.86 N 126.60 W
				H = 05 19 35.3 h = normal MB=5.0 MS=4.9
				D = 80.91 Az = 25.4 (NEIS)
				PV A 2.2s 65.5nm M = 5.2
12.	eP	A	06 11 11	Southern Sinkiang Prov., China
	LmH	B	30.4	39.20 N 74.21 E
	LmV	B	31.4	H = 06 03 00.2 h = normal MB = 5.2
				D = 44.51 Az = 306.3 (NEIS)
				PV A 1.4s 14.0nm M = 4.6
				LmH B 17 0.9/um 4.8
				LmV B 16 0.9/um 4.9
12.	eP	A	06 35 59	Tadzhik-Sinkiang Border Region
				39.37 N 73.84 E
				H = 06 27 51.1 h = normal MB = 4.8
				D = 44.18 Az = 306.1 (NEIS)
12.	ePKHKP	A	20 06 35	Fiji Islands Region 21.01 S 179.11 W
	ePKP2	A	06 42	H = 19 47 53.2 h = 606 km MB = 4.8
				D = 149.24 Az = 346.6 (NEIS)
				PKHKPV A 1.2s 8.1nm
12.	eP	A	20 27 14	El Salvador 13.56 N 89.89 W
	LmH	C	21 04.7	H = 20 14 37.3 h = 85 km MB = 5.0
	LmV	B	09.4	D = 86.78 Az = 38.7 (NEIS)
				LmH C 20s 0.6/um
				LmV B 16 0.5/um

Day	Phase		h m s	Remarks
12.	e(Pn)	A	20 52 26.5	<u>Czechoslovakia</u> 50.4 N 13.7 E
	e(Sg)	A	53 06	H = 20 51 53 h = 0 km
				D = 1.37 Az = 281 (ISC)
12.	eP	A	23 11 56.5	<u>Afghanistan</u> 35.0 N 68.6 E
				H = 23 03 53 h = 33 km MB = 4.7
				D = 43.52 Az = 309 (ISC)
				PV A 1.0s 11.8nm M = 4.8
12.	LmH	C	23 58.0	<u>South Sandwich Islands Region</u>
	LmV	C	58.0	58.75 S 25.11 W
				H = 22 56 20.8 h = 20 km MB = 5.1 (ISC)
				D = 113.0
				LmH C 20s 0.25 µm M = 4.9
				LmV C 20 0.55 µm 5.2
13.	+iP	A	03 10 46.5	<u>Eastern Kazakh SSR</u> 49.82 N 78.09 E
	ePn	A	12 18.5	H = 03 02 57.8 h = 0 km MB = 5.2
				D = 41.23 Az = 297.7 (NEIS)
				Underground explosion, MB = 5.9 (UPP)
				PV A 0.8s 42.3nm M = 5.2
13.	eP	A	04 59 59	<u>Turkey</u> 37.52 N 36.21 E
				H = 04 55 04.1 h = normal MB = 4.3
				D = 21.88 Az = 314.7 (NEIS)
				PV A 1.1s 12.1nm M = 4.2
13.	+iP	AB	08 04 20.0	<u>Near East Coast of Kamchatka</u>
	ePP	C	06 55	55.29 N 161.97 E
	eS	B	13 36	H = 07 53 02.7 h = 55 km MB = 5.8
	LmH	B	35.2	D = 71.58 Az = 340.6 (NEIS)
	LmV	B	35.8	PV A 1.3s 288.0nm M = 6.0
				LmH B 21.5 3.3 µm 5.6
				LmV B 22 2.4 µm 5.5
13.	eP	A	18 28 08	<u>Greece</u> 40.52 N 23.43 E
	LmH	B	32.6	H = 18 24 59.2 h = 24 km MB = 4.5
	LmV	B	33.5	D = 13.06 Az = 324.8 (NEIS)
				LmH B 10s 1.5 µm M = 4.3
				LmV B 9.5 0.9 µm

Day	Phase		h m s	Remarks
14.	eP	A	02 50 46	<u>Kurile Islands</u> 49.35 N 153.45 E
	sP	A	51 44.5	H = 02 39 22.5 h = 180 km MB = 4.8
				D = 75.16 Az = 336.0 (NEIS)
				h = 168 km
				PV A 1.2s 16.3nm M = 4.6
14.	e	A	06 51 40	<u>Corsica</u> 43.34 N 8.17 E
	eSg	A	52 55	H = 06 48 46.4 h = 17 km MB = 4.1
				D = 7.68 Az = 16.6 (NEIS)
14.	ePKHKP	A	08 56 33.5	<u>Fiji Islands Region</u> 19.74 S 178.26 W
	ePKP2	A	56 38	H = 08 37 53.3 h = 602 km MB = 4.4
				D = 148.18 Az = 348.1 (NEIS)
15.	ePKP2	A	08 37 08	<u>South of Fiji</u> 37.63 S 178.13 E
				H = 08 16 20.3 h = 61 km MB = 5.1 (NEIS)
				D = 163.82
15.	ePKP	A	20 53 22.5	<u>New Hebrides Islands</u> 18.60 S 169.20 E
				H = 20 34 18.8 h = 244 km MB = 5.1
				D = 143.39 Az = 336.0 (NEIS)
				PKPV A 1.3s 24.0nm
16.	eP	A	00 51 31.5	<u>Jujuy Province, Argentina</u>
				23.94 S 65.55 W
				H = 00 38 15.3 h = 280 km MB = 5.6
				D = 100.47 Az = 39.1 (NEIS)
				PV A 1.4s 18.6nm M = 5.4
16.	eP	A	16 53 59	<u>Tadzhik-Sinkiang Border Region</u>
	ePP	A	55 43.5	39.47 N 73.52 E
	LmH	C	17 14.1	H = 16 45 57.1 h = 64 km MB = 5.0
	LmV	C	14.1	D = 43.92 Az = 305.9 (NEIS)
				PV A 1.3s 21.8nm M = 4.8
				PPV A 1.3 17.5nm 4.8
				LmH C 18 1.0 μm
				LmV C 16 1.2 μm

Day	Phase		h m s	Remarks
16.	+iP	A	21 08 58.6	Kurile Islands 44.32 N 148.74 E
	LmH	B	48.0	H = 20 57 03.3 h = 54 km MB = 5.3
	LmV	B	48.0	D = 78.34 Az = 333.7 (NEIS)
				PV A 1.3s 48.0nm M = 5.3
				LmH B 16 0.9 µm
				LmV B 17 1.0 µm
16.	-eP	A	22 07 31.5	Kurile Islands 49.63 N 155.94 E
	eX	A	07 36	H = 21 55 50.8 h = 48 km MB=5.5 MS=4.7
	LmH	B	45.0	D = 75.56 Az = 337.5 (NEIS)
	LmV	B	45.0	PV A 1.2s 40.7nm M = 5.2
				XV A 1.6 41.2nm
				LmH B 17 1.0 µm 5.2
				LmV B 18 0.9 µm 5.2
17.	eP	A	02 12 47.5	Kodiak Island Region 56.72 N 151.66 W
	e	A	12 53	H = 02 01 23.2 h = 17 km MB=5.0 MS=5.1
	LmH	B	47.5	D = 72.11 Az = 11.1 (NEIS)
	LmV	B	51.8	PV A 1.6s 55.0nm M = 5.4
				LmH B 17.5 0.8 µm 5.1
				LmV B 18 0.8 µm 5.1
17.	eP	A	04 21 09	Greece - Albania Border Region
				40.34 N 20.63 E
				H = 04 18 12.2 h = 47 km MB = 4.4
				D = 12.08 Az = 331.6 (NEIS)
17.	eP	A	05 13 25	Greece - Albania Border Region
	LmH	B	17.6	40.29 N 20.64 E
	LmV	B	18.7	H = 05 10 31.5 h = 17 km MB=5.2 MS=5.0
				D = 12.12 Az = 331.6 (NEIS)
				PV A 1.4s 27.9nm M = 5.3
				LmH B 14 8.4 µm 4.9
				LmV B 14 4.6 µm
17.	eP	A	17 42 25	Off East Coast of Kamchatka
	epP	A	42 40	52.25 N 160.27 E
				H = 17 30 52.6 h = 57 km MB = 4.6
				D = 74.11 Az = 339.9 (NEIS)
				h = 57 km

Day	Phase		h m s	Remarks
18.	eP	A	05 25 05.5	Crete 34.48 N 24.26 E
	e	A	25 16	H = 05 20 40.8 h = 20 km MB = 4.2
				D = 18.60 Az = 334.1 (NEIS)
18.	eP	A	09 10 03.5	Greece - Albania Border Region
				40.24 N 20.62 E
				H = 09 07 02.8 h = 10 km MB = 4.7
				D = 12.17 Az = 331.8 (NEIS)
18.	ePP	A	18 44 40	Java 7.47 S 107.26 E
	e	A	44 54.5	H = 18 27 20.5 h = 88 km MB = 5.2 (ISC)
				D = 99.3
19.	eP	A	02 01 16	Kurile Islands 43.37 N 147.35 E
				H = 01 49 14.6 h = normal MB = 4.7
				D = 78.75 Az = 333.1 (NEIS)
19.	eP	A	20 26 26	Ascension Island Region
	LmH	B	54.6	10.27 S 13.09 W
	LmV	B	55.7	H = 20 15 52.6 h = normal MB = 4.8
				D = 64.37 Az = 17.2 (NEIS)
				PV A 1.9s 30.3nm M = 5.1
				LmH B 16 0.7 um 4.9
				LmV B 16 0.7 um 5.0
20.	eP	AB	01 04 58	Hokkaido, Japan Region
	eS	C	14 50	42.79 N 144.95 E
	LmH	B	40.3	H = 00 53 01.0 h = 51 km MB = 5.6
	LmV	B	42.9	D = 78.44 Az = 331.8 (NEIS)
				PV A 1.8s 112.0nm M = 5.5
				LmH B 18.5 1.7 um
				LmV B 19 1.8 um
20.	ePKHKP	A	19 44 19	Tonga Islands Region 23.77 S 175.91 W
	e	A	47 03.5	H = 19 24 24.3 h = normal MB=5.3 MS=5.4
				D = 152.54 Az = 349.6 (NEIS)
				PKHKPV A 1.4s 30.2nm

Day	Phase		h m s	Remarks
20.	ePKP2	A	20 09 29.5	South Island, New Zealand 44.35 S 168.12 E H = 19 48 37.7 h = 12 km MB = 5.4 D = 162.96 Az = 300 (ISC) PKP2V A 1.1s 16.1nm
20.	ePKIKP LmH LmV	A B B	21 38 55 22 22.8 30.7	East New Guinea Region 6.24 S 146.06 E H = 21 20 12.3 h = 111 km MB = 5.8 D = 121.77 Az = 327.7 (NEIS) PKIKPV A 1.0s 49.2nm LmH B 23 3.7/um LmV B 18 1.3/um
20.	eP	A	21 49 00	
21.	eP	A	01 15 08.5	Afghanistan-USSR Border Region 37.66 N 71.68 E H = 01 07 08.9 h = 93 km MB = 4.9 D = 43.84 Az = 307.2 (NEIS) traces
21.	e ePP LmH LmV	A A C C	03 31 56 32 30 04 20.0 22.5	Banda Sea 6.42 S 129.00 E H = 03 13 05.6 h = normal MB=5.4 MS=5.5 D = 112.13 Az = 322.4 (NEIS) LmH C 24s 0.9/um M = 5.3 LmV C 24 1.0/um 5.3
21.	ePKIKP ePKP2	A A	06 15 53 16 45.5	South Island, New Zealand 44.37 S 168.14 E H = 05 55 57.6 h = 49 km MB=5.8 MS=5.3 D = 162.98 Az = 299.9 (NEIS) PKP2V A 2.0s 154.0nm
21.	eP epP	A A	06 38 15 39 29	India-East Pakistan Border Region 25.68 N 90.91 E H = 06 27 41.8 h = 27 km MB = 4.7 D = 63.95 Az = 315.9 (NEIS) h = 47 km PKP2V A 2.0s 154.0nm

Day	Phase		h m s	Remarks
21.	ePKHKP	A	11 42 09	South of Fiji Islands 23.69 S 176.44 W
	ePKP2	A	42 18.5	H = 11 22 37.3 h = 241 km MB = 4.5
				D = 152.38 Az = 348.9 (NEIS)
21.	ePKIKP	AC	13 00 12.5	South of Fiji Islands 23.71 S 175.98 W
	ePKHKP	A	00 19.5	H = 12 40 22.1 h = normal MB=5.6 MS=6.3
	ePKP2	A	00 30.5	D = 152.47 Az = 349.5 (NEIS)
	ePP	B	03 50	LmH B 20s 7.9/um M = 6.4
	LmH	B	14 17.0	LmV B 20 8.0/um 6.5
	LmV	B	18.1	
21.	eP	A	16 06 20	Kamchatka 52.17 N 157.51 E
	epP	A	06 51.5	H = 15 54 59.2 h = 118 km MB = 5.8
	e	A	07 38.5	D = 73.57 Az = 338.1 (NEIS)
				h = 129 km
				PV A 1.7s 97.0nm M = 5.3
21.	eP	A	16 16 10	Svalbard Region 76.47 N 7.10 E
				H = 16 10 39.1 h = normal MB = 4.7
				D = 25.98 Az = 173.4 (NEIS)
				PV A 1.8s 33.8nm M = 4.6
21.	ePKHKP	A	19 45 33	South of Fiji Islands 22.38 S 179.69 E
	ePKP2	A	45 42	H = 19 26 45.7 h = 576 km MB = 5.1
	e	A	45 50	D = 150.28 Az = 344.6 (NEIS)
				PKHKPV A 1.1s 40.3nm
21.	ePKHKP	A	22 51 18	Fiji Region 21.77 S 178.79 W
				H = 22 32 30.8 h = 580 km MB = 4.5
				D = 150.05 Az = 347 (ISC)
22.	ePKHKP	A	04 30 52	South of Fiji Islands 23.76 S 176.06 W
				H = 04 10 58.2 h = normal MB=5.0 MS=4.8
				D = 152.50 Az = 349.4 (NEIS)
22.	ePKHKP	A	23 04 18	Tonga Islands Region 23.88 S 175.18 W
				H = 22 44 23.8 h = normal MB = 4.5
				D = 152.76 Az = 350.5 (NEIS)

Day	Phase		h m s	Remarks
23.	eP	A	15 01 53	Szechwan Province, China
				33.58 N 102.45 E
				H = 14 51 08.8 h = normal MB=5.1 MS=4.9
				D = 65.37 Az = 315.5 (NEIS)
23.	eP	A	16 01 02	Southern Greece 37.60 N 21.20 E
	e	A	01 14	H = 15 57 28.0 h = 49 km MB = 4.2
				D = 14.73 Az = 335.4 (NEIS)
23.	eP	A	19 37 16	Gabon 0.28 S 12.92 E
	ePP	C	39 10	H = 19 28 17.2 h = normal MB=5.9 MS=6.2
	eS	C	44 28	D = 50.75 Az = 358.9 (NEIS)
	eSS	C	48 15	PV A 2.0s 333.3nm M = 6.0
	LmV	B	20 01.4	LmH B 15 8.4 μm 5.9
	LmH	B	03.7	LmV B 15 8.5 μm 6.0
24.	eP	A	02 07 46	North Atlantic Ridge 13.61 N 44.84 W
				H = 01 57 46.1 h = normal MB=5.1 MS=4.5
				D = 58.51 Az = 38.5 (NEIS)
25.	LmH	C	16 53.0	East New Guinea Region 9.86 S 150.04 E
				H = 15 49 37.5 h = 10 km MB=5.4 MS=5.0
				D = 126.90 Az = 328.1 (NEIS)
				LmH C 34s 0.5 μm M = 5.0
26.	eP	A	07 42 36	Off East Coast of Honshu, Japan
				39.28 N 144.91 E
				H = 07 30 21.0 h = 42 km MB = 4.9
				D = 81.53 Az = 332.1 (NEIS)
26.	eP	A	11 41 06	Unimak Island Region 54.30 N 164.75 W
				H = 11 29 23.3 h = 31 km MB=4.9 MS=4.4
				D = 75.39 Az = 2.4 (NEIS)
26.	eP	A	15 17 18	Southern Nevada 37.13 N 116.07 W
	ePP	A	20 53	H = 15 05 00.2 h = 0 km MB=5.6 MS=4.2
				D = 81.23 Az = 30.7 (NEIS)
				Nuclear explosion (AEC)
				PV A 1.2s 32.5nm M = 5.3
				PPV A 1.2 20.33nm 5.2

Day	Phase		h m s	Remarks
27.	eP	AB	03 22 39.5	Off East Coast of Honshu, Japan
	Pm	A	23 03	33.60 N 141.13 E
	ePP	A	26 00	H = 03 10 07.9 h = 46 km MB=5.8 MS=6.1
	eS	B	33 00	D = 85.03 Az = 330.5 (NEIS)
	eSS	C	38 47	PV A 1.3s 50.2nm M = 5.5
	LmH	B	57.9	PmV A 1.5 226.1nm 6.1
	LmV	B	04 07.5	LmH B 18.5 16.1 µm 6.4
				LmV B 15 7.7 µm 6.2
27.	+eP	AB	04 21 26	Colombia 2.69 N 71.42 W
	eS	B	31 47	H = 04 09 01.3 h = 43 km MB=5.6 MS=5.8
				D = 83.51 Az = 39.5 (NEIS)
				PV A 2.3s 146.0nm M = 5.6
27.	-iP	A	05 36 30	Nepal 28.60 N 85.50 E
				H = 05 26 39.4 h = 70 km MB = 5.6
				D = 58.49 Az = 314.2 (NEIS)
				PV A 0.8s 88.5nm M = 5.9
27.	+iP	AB	05 59 27.5	Kurile Islands 43.18 N 146.65 E
	eS	B	06 09 22	H = 05 47 29.4 h = 43 km MB=6.0 MS=6.7
	LmH	B	39.0	D = 78.68 Az = 332.7 (NEIS)
	LmV	B	39.3	PV A 1.4s 326.0nm M = 6.1
				LmH B 17 101.0 µm 7.2
				LmV B 17 99.0 µm 7.3
27.	ePKIKP	A	24 02 44	South of Fiji Islands 25.29 S 178.57 E
	ePKHKP	A	02 51.5	H = 23 44 00.9 h = 605 km MB = 5.4
	iPKP2	A	03 05	D = 152.76 Az = 341.7 (NEIS)
				PKIKPV A 2.0s 42.7nm
				PKHKPV A 1.3 34.9nm
				PKP2V A 1.4 107.0nm
28.	ePP	A	01 39 21	Crete 34.97 N 23.97 E
	LmH	C	47.6	H = 01 35 00.5 h = 51 km MB = 4.5
	LmV	C	47.6	D = 18.06 Az = 333.9 (NEIS)
				PPV A 1.4s 14.0nm

Day	Phase		h m s	Remarks
28.	ePKP	A	07 30 55.5	Loyalty Islands Region 22.19 S 170.21 E
				H = 07 11 14.4 h = 31 km MB=5.4 MS=4.5
				D = 147.02 Az = 334.7 (NEIS)
				PKPV A 1.8s 50.6nm
28.	eP	A	21 06 08.5	South of Honshu, Japan 33.23 N 140.80 E
				H = 20 53 37.7 h = 59 km MB = 4.8
				D = 85.21 Az = 330.3 (NEIS)
29.	eP	A	06 39 57	Dodecanese Islands 35.50 N 27.88 E
	LmH	B	47.6	H = 06 35 34.4 h = 56 km MB = 4.6
	LmV	B	49.8	D = 19.18 Az = 327.1 (NEIS)
				PV A 1.1s 40.3nm M = 4.6
				LmH B 18 1.0 µm
				LmV B 14 0.6 µm
29.	eP	A	16 00 15	Southern Sinkiang Prov., China
	e(PP)	A	01 49.5	40.39 N 77.98 E
	eS	C	07 00	H = 15 51 51.7 h = normal MB=5.4 MS=5.1
	eSS	C	10 28	D = 46.16 Az = 306.0 (NEIS)
	LmH	B	20.7	PV A 1.6s 55.0nm M = 5.2
	LmV	B	20.7	LmH B 17 3.3 µm 5.4
				LmV B 16 4.3 µm 5.6
29.	eP	A	23 38 38.5	Tunisia 35.09 N 9.61 E
	LmV	B	46.2	H = 23 34 57.3 h = normal MB = 4.5
	LmH	B	46.3	D = 15.62 Az = 4.8 (NEIS)
				LmH B 13s 0.4 µm M = 3.1
				LmV B 14 0.6 µm
29.	eP	A	24 00 36	Sea of Okhotsk 53.48 N 153.68 E
				H = 23 50 03.9 h = 470 km MB = 4.6
				D = 71.45 Az = 335.6 (NEIS)
30.	eP	A	00 26 17.5	Near East Coast of Honshu, Japan
				38.51 N 142.05 E
				H = 00 14 05.7 h = 49 km MB = 4.7
				D = 81.13 Az = 330.6 (NEIS)

Day	Phase	h m s	Remarks
30.	eP A	07 51 04	Northern Sinkiang Prov., China 43.09 N 88.44 E H = 07 42 04.3 h = 32 km MB = 5.0 D = 50.86 Az = 306.9 (NEIS) PV A 1.1s 20.2nm M = 5.0
30.	eP A	08 28 58	Southern Sumatra 1.57 S 99.81 E H = 08 16 05 D = 90.05 Az = 320 (ISC)

Day	Phase		h m s	Remarks
1.	ePn	A	00 37 45.5	Southern Italy 39.69 N 18.77 E
	eSn	A	39 52	H = 00 34 40.2 h = 22 km MB = 3.9
	LmH	B	41.8	D = 12.06 Az = 337.7 (NEIS)
	LmV	B	42.0	LmH B 15.5s 2.0/um M = 4.2
				LmV B 12 0.5/um
1.	eP	C	04 21 25	South of Mariana Islands
	ePP	AB	25 37	12.03 N 141.12 E
	eSKS	C	32 06	H = 04 07 26.9 h = 43 km MB=5.3 MS=5.7
	ePS	C	34 42	D = 103.68 Az = 329.6 (NEIS)
	eSS	B	40 25	LmH B 17s 5.0/um M = 6.1
	LmH	B	05 13.0	LmV B 18 5.5/um 6.1
	LmV	B	13.0	
2.	eP	AC	03 08 30	Near Coast of Northern Peru
	ePP	AC	12 22	5.91 S 81.09 W
	eSKS	C	19 10	H = 02 54 59.7 h = 5 km MB=5.7 MS=5.6
	ePS	C	21 14	D = 96.24 Az = 39.8 (NEIS)
	eSS	C	26 15	PV A 2.2s 54.5nm M = 5.7
	LmH	B	57.2	PPV A 2.8 129.0nm 5.9
	LmV	B	58.4	LmH B 30 3.5/um 5.7
				LmV B 33 4.4/um 5.7
2.	ePKP2	A	14 26 21.5	Tonga Islands Region 22.74 S 175.31 W
				H = 14 06 29.0 h = normal MB=4.9 MS=4.8
				D = 151.62 Az = 350.7 (NEIS)
2.	eP	A	15 08 02.5	Near East Coast of Kamchatka
	LmV	C	42.0	51.67 N 158.08 E
				H = 14 56 30.1 h = 51 km MB = 5.2
				D = 74.17 Az = 338.6 (NEIS)
				PV A 1.0s 11.8nm M = 4.8
3.	eP	A	11 19 47	Hindu Kush Region 36.45 N 70.73 E
				H = 11 11 57.8 h = 198 km MB = 5.0
				D = 43.97 Az = 308.1 (NEIS)
				PV A 1.3s 17.5nm M = 4.4

Day	Phase		h m s	Remarks
3.	ePn	A	13 00 17.5	Probably explosion
	eSg	A	00 41.5	D c. 1.7
3.	eP	AB	14 35 12	Near Coast of Peru 12.27 S 77.80 W
	Pm	A	35 44	H = 14 21 29.1 h = 13 km MB=6.6 MS=7.6
	ePP	B	39 32	D = 99.00 Az = 40.1 (NEIS)
	eS	B	47 12	PV A 3.0s 500.0nm M = 6.5
	eSS	B	53 10	PV B 16 2.3 µm 6.4
	LmH	B	23.5	PmV A 3.5 2666.7nm 7.2
				PPV B 12 4.0 µm 6.7
4.	eP	A	04 11 13	Salta Province, Argentina
				22.75 S 63.73 W
				H = 03 58 31.0 h = 533 km MB = 5.0
				D = 98.50 Az = 38.5 (NEIS)
4.	eP	A	06 55 44	Off East Coast of Kamchatka
				52.29 N 160.26 E
				H = 06 44 09.5 h = normal MB=4.7 MS=3.9
				D = 74.06 Az = 339.8 (NEIS)
4.	eP	A	17 47 44	Off East Coast of Kamchatka
	e	A	47 58	52.67 N 159.00 E
				H = 17 36 13.7 h = normal MB=4.9 MS=4.1
				D = 73.44 Az = 339.0 (NEIS)
				PV A 1.0s 15.8nm M = 4.9
4.	eP	A	18 07 13	Off East Coast of Kamchatka
				52.69 N 159.27 E
				H = 17 55 42.4 h = normal MB=4.9 MS=4.4
				D = 73.48 Az = 339.2 (NEIS)
4.	eP	A	18 09 44	Off East Coast of Kamchatka
	LmH	B	44.3	52.41 N 159.96 E
	LmV	B	46.1	H = 17 58 10.7 h = normal MB = 5.0
				D = 73.89 Az = 339.6 (NEIS)
				PV A 1.0s 19.7nm M = 5.1
				LmH B 14 0.6 µm 5.0
				LmV B 12 0.3 µm 4.9

Day	Phase		h m s	Remarks
4.	-iP	AB	22 33 11.5	**West Pakistan** 26.29 N 66.54 E
	ePP	B	35 09	H = 22 24 32.7 h = normal MB=5.8 MS=5.9
	eS	B	40 06	D = 48.08 Az = 315.5 (NEIS).
	LmH	B	56.1	PV A 1.6s 560.4nm M = 6.3
	LmV	B	23 00.2	PV B 8 2.1 µm 6.0
				PPV B 5.5 1.2 µm 6.0
				LmH B 19 15.6 µm 6.0
				LmV B 14 8.2 µm 5.9
4.	+eP	AZN	22 47 03.5	**Off East Coast of Kamchatka**
	e	AZ	47 17.5	52.20 N 160.50 E
				H = 22 35 26.6 h = 16 km MB = 5.3
				D = 74.20 Az = 340.0 (NEIS)
				PV A 1.2s 40.6nm M = 5.3
4.	eP	A	23 24 29.5	**West Pakistan** 26.32 N 66.75 E
				H = 23 15 49.2 h = normal MB = 4.8
				D = 48.18 Az = 315.5 (NEIS)
5.	ePKP	A	03 21 08	**New Hebrides Islands** 20.40 S 169.59 E
				H = 03 01 42.3 h = 109 km MB = 4.6
				D = 145.16 Az = 335.3 (NEIS)
5.	eP	A	04 45 35	**Off East Coast of Kamchatka**
				52.24 N 160.23 E
				H = 04 34 00.3 h = normal MB = 4.6
				D = 74.11 Az = 339.8 (NEIS)
6.	eP	A	01 30 28	**Southern Persia** 26 1/2 N 56 E
				H = 01 22 43 (BCIS)
				D = 41.4
				PV A traces
6.	LmV	C	18 02.0	**Near Coast of Guerrero, Mexico**
	LmH	C	02.4	17.0 N 100.1 W
				H = 17 09 39 h = 51 km MB = 5.1 (ISC)
				D = 90.0
				LmH C 22s 0.3 µm
				LmV C 22s 0.3 µm

Day	Phase		h m s	Remarks
6.	e(P)	A	20 05 39	South of Honshu, Japan 33.17 N 140.63 E H = 19 53 02.7 h = 67 km MB = 4.6 D = 85.20 Az = 330.2 (NEIS)
7.	LmH	B	05 54.0	Near Coast of Peru 12.17 S 77.8 W
	LmV	B	56.7	H = 04 57 55 h = 55 km MB = 5.0 (ISC) D = 98.9 LmH B 19s 0.5/um LmV B 18 0.5/um
7.	+eP	A	10 04 18	Hokkaido, Japan Region 45.93 N 143.14 E H = 09 53 09.7 h = 325 km MB = 5.2 D = 75.07 Az = 330.4 (NEIS) PV A 1.3s 39.4nm M = 5.0
7.	eP	A	11 46 30	Southern Italy 39.73 N 18.85 E
	LmH	B	50.8	H = 11 43 36.9 h = 19 km MB = 4.6
	LmV	B	52.0	D = 12.04 Az = 337.4 (NEIS) PV A 0.9s 15.6nm M = 5.3 LmH B 14 3.3/um 4.5
7.	+eP	A	17 22 51	Nicaragua 11.58 N 85.44 W
	ePP	A	26 20	H = 17 10 36.3 h = 225 km MB = 5.1 D = 85.58 Az = 39.3 (NEIS) PV A 1.0s 39.4nm M = 5.2
7.	ePKIKP	A	22 11 16	South Sandwich Islands Region
	e	A	14 24	58.30 S 27.37 W
	ePS	C	21 14	H = 21 52 40.4 h = normal MB = 6.0
	ePKKP	A	21 37	D = 113.18 Az = 25.8 (NEIS)
	eSS	C	27 40	PKKPV A 3.0s 210.5nm
8.	eP	A	03 16 42	Central Siberia 60.59 N 118.15 E H = 03 07 14.9 h = normal MB = 4.6 D = 54.47 Az = 311.4 (NEIS)
8.	-iP	A	10 01 44.0	Leeward Islands 17.34 N 62.01 W
	ePP	B	04 16	H = 09 50 58.1 h = 47 km MB=6.6 MS=7.5
	eS	B	10 28	D = 66.43 Az = 41.8 (NEIS)

Day	Phase		h m s	Remarks
cont. 8.	eSKS	B	10 11 36	PV A 2.1s 1150.0nm M = 6.6
	LmV	B	26.7	PV B 16 35.4 /um 7.1
	LmH	B	26.8	LmH B 20 74.7 /um 6.9
				LmV B 20 88.6 /um 7.1
8.	eP	A	19 54 25	Honshu, Japan 36.15 N 139.76 E
				H = 19 42 08.7 h = 56 km MB = 5.0
				D = 82.27 Az = 329.6 (NEIS)
9.	ePKP	A	06 56 37	Fiji Islands Region 18.77 S 177.12 W
				H = 06 37 36.4 h = 394 km MB = 4.7
				D = 147.45 Az = 349.7 (NEIS)
9.	+iP	AB	07 43 59.5	Kurile Islands Region 44.72 N 150.12 E
	eS	B	53 45	H = 07 32 02.2 h = 49 km MB=6.3 MS=6.4
	eSS	C	59 00	D = 78.41 Az = 384.5 (NEIS)
	eP'P'	A	08 11 08	PV A 1.4s 686.0nm M = 6.5
	e	A	11 23	LmH B 20 45.1 /um 6.8
	LmH	B	08 21.8	LmV B 19 53.6 /um 6.9
	LmV	B	22.0	
9.	eP	A	18 24 09	Ionian Sea 37.57 N 20.81 E
				H = 18 20 43.4 h = 40 km MB = 4.0
				D = 14.63 Az = 336.2 (NEIS)
9.	ePKP	A	20 44 55.5	Tonga 20.5 S 173.2 W
				H = 20 25 12 h = 33 km MB = 4.7
				D = 149.65 Az = 354 (ISC)
10.	ePKHKP	A	02 05 59	South of Fiji Islands
	ePKP2	A	06 07	22.12 S 179.49 W
				H = 01 47 12.7 h = 588 km MB = 5.1
				D = 150.23 Az = 345.7 (NEIS)
				PKHKPV A 1.4s 34.9nm
10.	epPKP	A	02 53 19	South of Fiji Islands
				26.50 S 177.12 W
				H = 02 33 13.3 h = 97 km MB = 5.1
				D = 154.98 Az = 346.8 (NEIS)

Day	Phase		h m s	Remarks
10.	eiPn	A	04 04 10.5	<u>Austria</u> 47.86 N 12.28 E
	ePg	A	04 22	H = 04 03 27.4 h = normal
	eSn	A	04 48	D = 2.83 Az = 351.4 (NEIS)
	eSg	A	05 03.5	
10.	ePn	A	05 17 21	<u>Austria</u> 47.5 N 12.5 E
	eiSg	A	18 13	H = 05 16 32 h = 9 km
				D = 3.19 Az = 350 (ISC)
10.	+eP	AB	07 00 19	<u>Hokkaido, Japan Region</u>
	ePP	AB	03 18	40.96 N 143.13 E
	eS	B	10 16	H = 06 48 14.0 h = 29 km
				MB = 5.8 MS = 6.2 (NEIS)
				D = 79.40
				PV A 2.0s 256.4nm M = 5.9
				PPV A 1.6 87.9nm 5.7
10.	eP1	AB	07 08 51.5	<u>Off East Coast of Honshu, Japan</u>
	+iP2	A	08 56.5	40.93 N 143.11 E
	ePP	B	11 52	H = 06 56 49.0 h = 45 km MB=5.8 MS=6.2
	eS	B	18 54	D = 79.42 Az = 331.0 (NEIS)
	LmH	B	42.8	P1V A 1.3s 34.9nm M = 5.2
	LmV	B	47.6	P2V A 2.0 461.5nm 6.1
				PPV A 1.6 126.4nm 5.8
				LmH B 18 56.8 μm 7.0
				LmV B 18 51.6 μm 6.9
10.	eP	A	16 17 33.5	<u>Afghanistan-USSR Border Region</u>
				37.46 N 71.81 E
				H = 16 09 35.7 h = 119 km MB = 4.8
				D = 44.0 Az = 307.4 (NEIS)
				PV A 0.8s 19.2nm M = 4.9
10.	eP	A	20 06 38.5	<u>Near Coast of Peru</u> 12.44 S 77.61 W
	LmV	B	55.8	H = 19 52 59.0 h = 27 km MB=5.3 MS=5.1
	LmH	B	56.3	D = 99.01 Az = 40.1 (NEIS)
				PV A 1.6s 11.0nm M = 5.1
				LmH B 16 0.25 μm 4.8
				LmV B 16 0.35 μm 5.0

Day	Phase		h m s	Remarks
10.	eP	A	21 45 28.5	**Southern Sumatra** 4.14 S 102.83 E
	e	A	45 49	H = 21 32 10.6 h = 21 km MB = 6.0
	LmH	B	22 50.9	D = 93.94 Az = 320.4 (NEIS)
	LmV	B	50.9	PV A 1.2s 48.7nm M = 5.8
				LmV B 18 0.3 µm 4.8
10.	LmH	B	22 00.7	**Off Coast of Peru** 13.04 S 77.38 W
	LmV	B	00.7	H = 20 57 51 h = 62 km MB = 4.7 (ISC)
				D = 99.3
				LmH B 15s 0.3 µm
				LmV B 16 0.35 µm
11.	ePKHKP	A	06 20 20	**West Macquarie Island** 60.66 S 153.32 E
	LmH	B	07 40.8	H = 06 00 14.2 h = normal MB = 5.3
	LmV	B	41.3	D = 156.52 Az = 262.0 (NEIS)
				LmH B 18s 0.6 µm M = 5.4
				LmV B 19 0.7 µm 5.5
11.	e	A	08 54 10	**West of Macquarie Island**
	ePKP2	A	54 18	60.75 S 153.93 E
	e	A	54 27.5	H = 08 33 52.3 h = normal MB=5.3 MS=6.1
	LmV	V	10 14.6	D = 156.81 Az = 261.2 (NEIS)
	LmH	B	16.7	LmH B 18.5s 3.3 µm M = 6.1
				LmV B 18 2.8 µm 6.1
11.	eP	AB	09 17 22	**Iceland Region** 67.65 N 20.10 W
				H = 09 12 19.3 h = normal MB=4.8 MS=5.4
				D = 23.13 Az = 121.5 (NEIS)
				PV A 1.2s 40.6nm M = 4.8
11.	eP	A	14 24 08	**North of Severnaya Zemlya**
				83.46 N 117.73 E
				H = 13 41 21.4 h = normal MB = 4.3
				D = 41.77 Az = 293.3 (NEIS)
				PV A 1.2s 12.2nm M = 4.5

Day	Phase		h m s	Remarks
11.	+eP	A	14 24 23	Taiwan 23.15 N 121.37 E
	LmH	B	15 06.9	H = 14 11 56.1 h = 44 km MB = 5.3
	LmV	B	06.9	D = 84.06 Az = 323.0 (NEIS)
				PV A 1.4s 37.2nm M = 5.2
				LmH B 14 2.1/um 5.7
				LmV B 14 3.4/um 5.9
11.	eP	A	18 22 57.5	Northwest of Kurile Islands
				52.30 N 152.40 E
				H = 18 12 16.5 h = 421 km MB = 4.5
				D = 72.20 Az = 335.0 (NEIS)
				PV A 1.0s 15.8nm M = 4.6
12.	+eP	AB	04 59 39.5	Off East Coast of Honshu, Japan
	eS	C	05 09 40	40.49 N 143.49 E
	LmH	B	33.3	H = 04 47 31.4 h = 26 km MB=5.3 MS=5.3
	LmV	B	39.3	D = 79.94 Az = 331.2 (NEIS)
				PV A 1.7s 48.5nm M = 5.2
				LmH B 18.5 3.5/um 5.7
				LmV B 17 3.2/um 5.8
12.	eP	A	05 26 37	North Atlantic Ridge 49.69 N 29.42 W
				H = 05 21 01.1 h = normal MB = 4.4
				D = 26.07 Az = 72.0 (NEIS)
				PV A traces
12.	+eP	AB	06 27 00	Off East Coast of Honshu, Japan
	eS	BC	36 48	40.51 N 143.58 E
	eSS	C	42 00	H = 06 14 51.5 h = 24 km MB=5.5 MS=6.0
	LmH	B	07 00.9	D = 79.96 Az = 331.3 (NEIS)
	LmV	B	13.5	PV A 2.4s 269.0nm M = 5.8
				LmH B 19 17.3/um 6.5
				LmV B 13.5 9.9/um 6.4
12.	eP	A	12 44 56	Kodiak Island Region 56.15 N 153.72 W
				H = 12 33 24.9 h = 10 km MB=4.8 MS=4.8
				D = 72.88 Az = 9.7 (NEIS)
				PV A 2.2s 109.1nm M = 5.6

Day	Phase		h m s	Remarks
12.	eP	A	12 57 37.5	Off East Coast of Honshu, Japan
	e	A	57 48	40.37 N 143.61 E
	LmH	B	13 36.6	H = 12 45 25.7 h = 10 km MB = 4.7
	LmV	B	40.1	D = 80.09 Az = 331.3 (NEIS)
				PV A 1.6s 22.0nm M = 4.9
				LmH B 14.5 0.5 µm 5.0
				LmV B 15 0.5 µm 4.9
12.	eP	A	16 29 33.5	Off East Coast of Honshu, Japan
	e	A	29 43	40.31 N 143.69 E
	LmH	B	17 08.7	H = 16 17 23.1 h = 14 km MB = 4.9
	LmV	B	12.5	D = 80.18 Az = 331.3 (NEIS)
				PV A 2.2s 43.6nm M = 5.1
				LmH B 13 0.6 µm 5.1
				LmV B 13 0.5 µm 5.1
12.	ePKP	A	19 18 18	Fiji Islands Region 17.88 S 178.68 W
				H = 18 59 44.2 h = 623 km MB = 4.9
				D = 146.29 Az = 348.2 (NEIS)
				PKPV A 1.4s 14.0nm
12.	ePKP2	A	22 05 56	Tonga 20.29 S 174.07 W
				H = 21 46 05.2 h = 33 km MB = 4.8 (NEIS)
				D = 149.38
				traces
13.	eP	A	02 48 57	Molucca Passage 0.52 N 125.99 E
	epP	A	49 04.5	H = 02 34 53.1 h = 41 km MB=5.5 MS=5.1
	ePP	A	53 15	D = 104.82 Az = 323.1 (NEIS)
	epPP	A	53 21	PV A traces
	LmV	B	03 38.7	LmH B 22s 0.7 µm M = 5.2
	LmH	B	38.8	LmV B 20 0.9 µm 5.3
13.	ePKP2	A	07 15 58	Tonga Islands 20.31 S 173.99 W
				H = 06 56 06.7 h = normal MB=4.3 MS=4.7
				D = 149.41 Az = 353.0 (NEIS)
				traces

Day	Phase		h m s	Remarks
13.	LmH	B	22 02.3	**Tibet** 34.76 N 87.23 E
	LmV	B	06.1	H = 21 29 52.2 h = 33 km MB = 5.1 (ISC)
				D = 55.5
				LmH B 16s 0.9 /um M = 4.9
				LmV B 16 0.6 /um 4.8
14.	eP	A	04 13 13.5	**Yugoslavia** 43.72 N 17.58 E
	e	A	15 09	H = 04 11 16.3 h = normal MB = 4.2
				D = 8.03 Az = 331.7 (NEIS)
14.	eP	A	07 58 04	**Peru-Ecuador Border Region**
				2.38 S 76.45 W
				H = 07 45 17.2 h = 162 km MB = 5.1
				D = 90.59 Az = 39.5 (NEIS)
				PV A 1.7s 30.3nm M = 5.0
14.	eP	AB	14 23 51.5	**Off East Coast of Honshu, Japan**
	ePP	BC	26 52	40.57 N 143.68 E
	eS	B	33 50	H = 14 11 41.1 h = 15 km MB=5.3 MS=5.7
	LmH	B	58.0	D = 79.94 Az = 331.3 (NEIS)
	LmV	B	15 05.8	PV A 2.0s 106.8nm M = 5.5
				LmH B 19 10.3 /um 6.2
				LmV B 15 4.4 /um 5.9
14.	eP	A	21 14 12.5	**Off East Coast of Honshu, Japan**
				40.54 N 143.68 E
				H = 21 02 03.0 h = 21 km MB = 4.9
				D = 79.97 Az = 331.3 (NEIS)
14.	ePKHKP	A	21 20 21	**South of Fiji Islands** 25.51 S 179.87 E
	ePKP	A	20 37.5	H = 21 01 17.7 h = 462 km MB = 4.8
				D = 153.33 Az = 343.2 (NEIS)
15.	+eP	AB	01 28 55.5	**Off East Coast of Honshu, Japan**
	eS	B	38 55	40.63 N 143.68 E
	LmH	B	02 03.2	H = 01 16 47.1 h = 22 km MB=5.4 MS=5.5
	LmV	B	09.6	D = 79.89 Az = 331.3 (NEIS)
				PV A 1.8s 67.5nm M = 5.4
				LmH B 18.5 6.1 /um 6.0
				LmV B 14 2.5 /um 5.7

Day	Phase		h m s	Remarks
15.	eP	A	06 16 04	**Kurile Islands** 44.45 N 148.37 E H = 06 04 11.9 h = 79 km MB = 5.0 D = 78.11 Az = 333.5 (NEIS) PV A 1.4s 23.3nm M = 4.9
15.	ePKP	A	07 09 10	**Samoa Islands Region** 16.58 S 172.71 W H = 06 49 32.2 h = normal MB=4.9 MS=5.0 D = 145.84 Az = 355.1 (NEIS) PKPV A 2.0s 34.2nm
	LmH	C	08 21.0	
	LmV	C	21.0	
15.	ePKP2	A	07 51 51.5	**Antarctica** 70.52 S 161.53 E H = 07 31 42.0 h = 33 km MB = 4.9 (NEIS) D = 155.68
15.	eP	A	09 59(57)	**Greece** 40.73 N 22.98 E H = 09 56 52.4 h = 33 km MB = 4.2 (NEIS) D = 12.69 PV A traces LmH C 14s 0.8/um M = 3.9
	LmH	C	10 04.1	
15.	eP	A	15 28 01	**Off East Coast of Kamchatka** 52.27 N 160.3 E H = 15 16 20.5 h = 0 km MB = 4.8 D = 74.09 Az = 340 (ISC)
15.	eP	A	16 13 43	**Off East Coast of Kamchatka** 52.56 N 159.33 E H = 16 02 12.1 h = normal MB = 4.7 D = 73.61 Az = 339.2 (NEIS)
15.	ePKIKP	A	21 47 34	**Kermadec Islands** 30.71 S 178.01 W H = 21 27 42.5 h = 59 km MB = 5.7 D = 158.82 Az = 342.9 (NEIS) PKIKPV A 1.8s 47.3nm PKP2V A 1.5 52.8nm LmH C 28 0.4/um LmV C 35 0.35/um
	ePKP2	A	48 09.5	
	LmV	C	22 50.5	
	LmH	C	51.5	

Day	Phase		h m s	Remarks
16.	ePn	A	03 42 52.5	Svabian Jura Region, Fed. Rep. of Germany
	eiPg	A	43 02.5	48.29 N 9.12 E
	eSn	A	43 26	H = 03 42 08.6 h = 21 km
	eSg	A	43 39	D = 2.87 Az = 33.7 (NEIS)
16.	eP	A	05 09 24	Southern Sinkiang Prov., China
				39.28 N 74.12 E
				H = 05 01 13.0 h = normal MB = 4.9 (NEIS)
				D = 44.41
16.	+eiP	A	05 42 05	North Atlantic Ocean 52.65 N 32.16 W
				H = 05 36 27.6 h = normal MB = 5.0 (NEIS)
				D = 26.93
				PV A 1.4s 34.9nm
16.	+iP1	AB	05 50 48.8	North Atlantic Ocean 52.64 N 32.07 W
	iP2	AB	51 00	H = 05 45 09.8 h = normal
	eiS	B	55 11	MB = 5.8 MS = 6.9 (NEIS)
	LmH	B	06 01.3	D = 26.88
	LmV	B	01.4	P1V A 1.0s 138.0nm M = 5.5
				P2V A 1.6 879.1nm 6.1
				P2V B 16 18.8 μm 6.5
				LmH B 16 535.4 μm 7.2
16.	+iP	A	06 40 49.3	Eastern Kazakh SSR 49.97 N 78.97 E
	ePn	A	42 20.5	H = 06 32 57.5 h = 0 km MB = 5.5 (NEIS)
	ePP	A	42 25.5	D = 41.66
				Underground explosion MB = 6.4 (UPP)
				PV A 1.0s 102.0nm M = 5.5
16.	+iP	AB	09 41 58.5	Off East Coast of Honshu, Japan
	e(PP)	A	45 06	40.34 N 143.67 E
	e	A	45 16	H = 09 29 49.0 h = 24 km
	LmH	B	10 18.6	MB = 5.6 MS = 5.2 (NEIS)
	LmV	B	24.8	D = 80.14
				PV A 1.6s 90.7nm M = 5.5
				LmH B 16 1.7 μm 5.5
				LmV B 13 1.4 μm 5.5

Day	Phase		h m s	Remarks
16.	ePKIKP	A	17 49 24.5	New Britain Region 6.32 S 148.38 E
	LmH	C	18 32.4	H = 17 30 34.7 h = 70 km MB = 5.7 (NEIS)
	LmV	C	41.0	D = 123.05
				LmH C 30s 0.7/um
				LmV C 24 0.4/um
16.	eSg	A	19 51 16.5	Poland 50.36 N 18.88 E
				H = 19 48 45.1 M = 3.2 (WAR)
				D = 4.7
17.	LmH	B	22 30.6	Afghanistan-USSR Border Region
	LmV	C	32.6	36.64 N 71.24 E
				H = 22 09 08.6 h = 191 km MB = 4.5 (ISC)
				D = 44.2
				LmH B 16s 0.5/um
				LmV C 14 0.6/um
18.	eP	A	00 37 29	Leeward Islands 17.60 N 62.21 W
	epP	A	37 41	H = 00 26 42.6 h = 45 km
	LmH	C	59.0	MB = 5.2 MS = 4.4 (NEIS)
	LmV	C	59.0	D = 66.36 h = 46 km
				PPV A 2.0s 42.7nm M = 5.5
				LmH C 30 0.6/um 4.7
				LmV C 30 0.9/um 4.9
18.	ePKIKP	A	09 22 48	Near North Coast of New Guinea
	e	A	23 23.5	3.16 S 142.04 E
	e	A	23 46	H = 09 04 04.3 h = 36 km MB = 5.4 (NEIS)
	ePP	A	23 54.5	D = 117.03
	eSS	C	40 10	LmH B 18s 0.6/um M = 5.3
	LmH	B	10 16.1	LmV B 17 0.7/um 5.4
	LmV	B	20.8	
18.	ePKP	AB	12 11 26	Samoa Islands Region
	e	A	11 30	16.33 S 172.45 W
	LmH	C	13 16.0	H = 11 51 49.0 h = normal
				MB = 5.4 MS = 5.2 (NEIS)
				D = 145.62
				PKPV A 1.5s 123.0nm

Day	Phase		h m s	Remarks
18.	eP	A	14 14 18.5	South Unimak Island Region 53.63 N 163.80 W
				H = 14 02 33.8 h = 42 km MB = 4.7 (NEIS)
				D = 76.03
				PV A 1.2s 16.3nm
18.	e	A	17 37 36.5	South Pacific Cordillera
	ePKP2	A	37 47	55.39 S 128.78 W
				H = 17 17 44.3 h = normal
				MB = 5.0 MS = 5.5 (NEIS)
				D = 155.95
				PKP2V A 1.3s 21.8nm
19.	ePKIKP	A	04 05 46.5	New Hebrides 14.21 S 167.13 E
				H = 03 46 36.0 h = 165 km MB = 4.8 (NEIS)
				D = 138.59
19.	e	A	04 09 15	South of Honshu, Japan 30.76 N 141.46 E
	ePP	A	12 35	H = 03 56 19.8 h = 27 km
	LmH	B	51.2	MB = 4.9 MS = 4.9 (NEIS)
	LmV	B	55.1	D = 87.63
				PPV A traces
				LmH B 15s 1.1/um M = 5.4
				LmV B 14 1.0/um 5.4
19.	eP	A	06 59 17.5	Northern Celebes 0.15 S 123.85 E
	epP	A	59 36	H = 06 45 17.0 h = 70 km MB = 5.5 (NEIS)
	ePP	A	07 03 40	D = 104.06 h = 62 km
	LmH	C	37.0	PV A 1.4s 16.3nm M = 5.7
				PPV A 1.4 23.3nm 5.6
				LmH C 38 2.10/um
19.	eP	A	09 52 04	Kurile Islands 47.27 N 152.02 E
				H = 09 40 31.0 h = 171 km MB = 4.4 (NEIS)
				D = 76.67
19.	e	A	11 10 18	

Day	Phase		h m s	Remarks
19.	eP	A	15 09 02	North Atlantic Ridge 31.00 N 41.45 W H = 15 00 58.4 h = normal MB = 4.5 MS = 4.6 (NEIS) D = 43.64 PV A 1.5s 22.6nm M = 4.7
20.	e	A	04 37 38	Tonga 20.27 S 174.39 W H = 04 17 35.3 h = 44 km MB = 5.0 (NEIS) D = 149.32
20.	ePKHKP ePKP2	A A	05 28 18 28 27	South of Fiji Islands 22.10 S 179.61 W H = 05 09 29.7 h = 570 km MB = 4.8 (NEIS) D = 150.19 PKHKPV A 1.2s 16.3nm
20.	ePn eSn i LmH LmV	A A A B B	11 28 46 31 08 32 09 33.1 34.2	Southern Italy 39.70 N 18.89 E H = 11 25 55.3 h = normal MB = 4.9 MS = 4.9 (NEIS) D = 12.08 LmH B 12.8s 9.5/um M = 5.1 LmV B 10 3.2/um
20.	eP	A	11 55 04	Hokkaido, Japan Region 42.32 N 142.31 E H = 11 43 05.1 h = 24 km MB = 5.3 MS = 4.9 (NEIS) D = 77.91 PV A 1.5s 30.2nm M = 5.1
20.	ePKIKP LmH LmV	A B B	15 46 39 16 45.1 45.4	Solomon Islands 6.60 S 154.79 E H = 15 27 39.6 h = 43 km MB = 5.6 MS = 5.7 (NEIS) D = 126.46 LmH B 20s 1.0/um M = 5.5 LmV B 20 0.7/um 5.4
20.	ePKIKP	A	19 58 31	Solomon Islands 6.59 S 154.72 E H = 19 39 31.1 h = 43 km MB = 5.5 (NEIS) D = 126.43

Day	Phase		h m s	Remarks
20.	LmH	B	20 26.9	LmH B 20s 1.3/um
	LmV	B	27.0	LmV B 20 1.5/um
21.	ePKHKP	A	02 34 44	South Pacific Cordillera
	LmV	B	03 40.8	54.32 S 133.10 W
	LmH	B	41.0	H = 02 14 39.3 h = 33 km MB = 5.4 (NEIS)
				D = 158.34
				LmH B 18s 0.8/um M = 5.5
				LmV B 20 1.2/um 5.7
21.	+ePKIKP	A	04 31 04	Fiji Islands Region 17.91 S 178.61 W
	ePKHKP	A	31 06	H = 04 12 29.4 h = 602 km MB = 6.0 (NEIS)
				D = 146.33
				PKIKPV A 1.2s 44.7nm
				PKHKPV A 1.3 140.0nm
21.	e(P)	A	04 41 45	
21.	+eiP	AB	12 59 39	Near East Coast of Kamchatka
	LmV	C	13 29.8	53.93 N 160.48 E
	LmH	C	30.3	H = 12 48 13.6 h = normal
				MB = 5.7 MS = 4.7 (NEIS)
				D = 72.57
				PV A 1.4s 209.0nm M = 5.9
				LmH C 29 1.1/um 5.0
				LmV C 28 0.9/um 4.9
21.	eP1	A	14 52 03	Southern Italy 39.73 N 15.05 E
	eP2	A	52 05.5	H = 14 49 29.4 h = 317 km MB = 4.0 (NEIS)
	e(pP)	A	53 21	D = 11.18
				P1V A 0.7s 5.8nm M = 4.0
				P2V A 1.1 26.2nm 4.4
21.	e	A	15 20 00	Off East Coast of Honshu, Japan
	LmH	C	53.8	40.56 N 143.58 E
	LmV	C	16 01.5	H = 15 07 41.0 h = normal MB = 4.7 (NEIS)
				D = 79.91
				LmH C 19s 1.1/um M = 5.2

Day	Phase		h m s	Remarks
22.	ePKP	A	01 51 41	Fiji Islands Region 17.00 S 177.69 W H = 01 32 10.0 h = normal MB = 4.4 (NEIS) D = 145.61
22.	eP	A	04 51 41	Iceland Region 62.34 N 26.02 W H = 04 46 32.6 h = 33 km MB = 4.0 (NEIS) D = 23.46
22.	eP eS LmV LmH	AB C B B	05 11 24 15 44 22.1 22.3	Iceland Region 62.12 N 26.38 W H = 05 06 16.2 h = normal MB = 5.1 MS = 5.4 (NEIS) D = 23.58 PV B 7s 1.0/um M = 5.4 LmH B 14 2.3/um 4.8 LmV B 15 3.3/um 5.1
22.	eP	A	06 17 50	North Atlantic Ridge 14.91 N 45.05 W H = 06 08 04 h = 50 km MB = 4.7 D = 57.63 Az = 39 (ISC)
22.	eP	A	09 17 37	Southern Sumatra 0.78 S 98.09 E H = 09 04 43.3 h = 33 km MB = 5.2 (NEIS) D = 88.35
22.	e(P)	A	09 29 34	Southern Sumatra 0.74 S 98.09 E H = 09 16 41.0 h = 84 km MB = 5.2 (NEIS) D = 88.32 (P)V A 1.9s 60.6nm
22.	eP1 eP2 +iP3 eS LmH LmV	A B A C B B	12 11 16.5 11 20 11 23.5 15 40 22.3 23.6	Iceland Region 62.09 N 26.17 W H = 12 06 11.2 h = normal MB = 4.9 MS = 5.6 (NEIS) D = 23.48 P1V A 0.9s 23.4nm M = 4.2 P2V B 7 1.6/um 5.6 P3V A 1.7 115.0nm 5.1 LmH B 13.5 3.2/um 4.9 LmV B 12 4.4/um 5.3

Day	Phase		h m s	Remarks
22.	ePKP	A	14 42 46.5	<u>Tonga</u> 17.46 S 174.62 W
				H = 14 23 21.4 h = 130 km MB = 4.8 (NEIS)
				D = 146.52
				PKPV A 1.1s 18.2nm
22.	eP	A	17 57 16.5	<u>Tadzhikistan</u> 37.55 N 72.25 E
				H = 17 49 18.3 h = 127 km MB = 4.8 (NEIS)
				D = 44.26
22.	eP	A	22 58 44.5	<u>Mindoro, Philippine Islands</u>
	ePP	A	23 02 26	13.48 N 120.57 E
	eSKS	C	09 14	H = 22 45 42.1 h = 41 km
	eS	C	15 25	MB = 5.2 MS = 5.1 (NEIS)
	LmH	B	41.5	D = 91.29
	LmV	B	45.6	PV A 1.6s 27.5nm M = 5.4
				LmH B 17 3.2 μm 5.8
				LmV B 16 3.2 μm 5.9
23.	ePKIKP1	A	06 33 55	<u>D'Entrecasteaux Islands Region</u>
	ePKIKP2	AB	34 04	8.42 S 154.03 E
	e	B	34 42	H = 06 14 54.0 h = 48 km MB=6.1 MS=7.2
	ePP	AB	36 06	D = 127.69 Az = 330.6 (NEIS)
	ePS	B	46 00	PKIKP1V A 2.0s 42.7nm
	eSS	B	53 10	PKIKP2V B 1.7 145.5 μm
	eSSS	B	58 05	PPV A 4 1677.0nm M = 6.6
	LmH	B	07 20.0	PPV B 15 8.7 μm 6.7
	LmV	B	30.5	LmH B 22.5 136.0 μm 7.6
				LmV B 21.5 74.3 μm 7.3
23.	eP	A	10 16 30.5	<u>North of Ascension Island</u> 0.73 S 16.14 W
	e	A	16 37.5	H = 10 06 48.5 h = 24 km MB = 4.7
				D = 56.39 AZ = 20.9 (NEIS)
23.	ePKIKP	A	11 46 21	<u>West of Macquarie Island</u> 54.18 S 143.73 E
				H = 11 26 36.2 h = normal MB = 5.6
				D = 151.02 Az = 282.9 (NEIS)

Day	Phase		h m s	Remarks
23.	eP	A	11 56 38	North of Ascension Island
				1.01 S 15.97 W
				H = 11 46 56.0 h = normal MB = 4.9 MS=5.1
				D = 56.58 Az = 20.7 (NEIS)
				PV A 1.3s 24.0nm M = 5.1
24.	eP	A	05 39 17	South of Honshu, Japan 33.41 N 140.82 E
	ePP	A	42 34.5	H = 05 26 47.0 h = 63 km MB = 5.4
				D = 85.06 Az = 330.3 (NEIS)
				PV A 1.6s 44.0nm M = 5.3
24.	e(P)	A	07 43 39	South of Honshu, Japan 30.89 N 141.55 E
	ePP	A	47 03.5	H = 07 30 49.5 h = 45 km MB=5.1 MS=4.8
	LmH	B	08 25.6	D = 87.56 Az = 330.7 (NEIS)
	LmV	B	29.7	LmH B 14s 1.0/um M = 5.4
				LmV B 14 1.1/um 5.4
24.	ePKP	A	21 21 59.5	Fiji Islands Region 17.42 S 178.67 W
				H = 21 03 20.7 h = 556 km MB = 5.0
				D = 145.84 Az = 348.3 (NEIS)
25.	eP	A	00 18 06.5	Near Coast of Chiapas, Mexico
	LmH	C	50.0	15.84 N 93.09 W
	LmV	C	50.5	H = 00 05 34.1 h = 120 km MB = 5.5
				D = 86.93 Az = 38.1 (NEIS)
				PV A 2.0s 59.8nm M = 5.2
				LmH C 27 0.8/um
				LmV C 27 1.9/um
25.	ePKIKP	A	03 38 07	New Britain Region 6.31 S 152.28 E
				H = 03 19 07.7 h = 18 km MB=5.7 MS=5.0
				D = 125.00 Az = 330.5 (NEIS)
				PKIKPV A 1.8s 33.8nm
25.	eP	A	04 59 55.5	North of Ascension Island
				0.91 S 16.15 W
				H = 04 50 14.7 h = normal MB = 4.8
				D = 56.56 Az = 20.8 (NEIS)

Day	Phase		h m s	Remarks
25.	eP	A	04 59 55.5	North of Ascension Island 0.91 S 16.15 W H = 04 50 14.7 h = normal MB = 4.8 D = 56.56 Az = 20.8 (NEIS)
25.	eP	A	11 47 45	Crete 34.65 N 23.42 E H = 11 43 35.6 h = 43 km MB = 4.9 D = 18.16 Az = 335.3 (NEIS) PV A 1.0s 27.6nm M = 4.4
25.	eP	A	15 52 18.5	Hokkaido, Japan Region 42.73 N 145.42 E H = 15 40 20.0 h = 43 km MB = 5.2 D = 78.66 Az = 332.1 (NEIS) PV A 1.5s 20.1nm M = 4.9
26.	eSn	A	15 10 42	Northern Italy 44.1 N 10.7 E
	eSg	A	11 29	H = 15 07 53 (BCIS) D = 6.60
26.	ePKP2	A	24 02 30	Kermadec Islands Region
	e	A	02 43	31.53 S 177.53 W H = 23 41 56.2 h = 50 km MB = 5.2 (NEIS) D = 159.74
27.	eP	A	02 40 52	Hokkaido, Japan Region 44.87 N 145.01 E H = 02 29 23.6 h = 223 km MB = 5.1 D = 76.64 Az = 331.6 (NEIS) PV A 1.3s 17.5nm M = 4.6
27.	ePKP	A	05 10 51	Loyalty Islands Region 21.76 S 170.57 E H = 04 51 18.5 h = 102 km MB = 5.1 D = 146.78 Az = 335.3 (NEIS) PKPV A 1.6s 27.4nm
27.	eP	A	08 56 40	Crete 34.27 N 25.14 E H = 08 52 20.4 h = 70 km MB = 4.4 D = 19.11 Az = 333.0 (NEIS) PV A 0.9s 19.5nm M = 4.4

Day	Phase		h m s	Remarks
28.	e	A	10 15 01.5	North of Ascension Island 0 S 13 W
				H = 10 05 23 MB = 4.3 (NORSAR)
				D = 24.6
29.	ePn	A	01 07 04.5	Yugoslavia 44.63 N 18.41 E
	eSn	A	08 27	H = 01 05 15.5 h = normal MB=5.1 MS=4.8
	eSg	A	09 20	D = 7.57 Az = 325.1 (NEIS)
	LmH	B	09.7	PnV A 1.4s 140.0nm M = 5.7
	LmV	B	10.3	LmH B 14 17.6 μm 4.9
				LmV B 11 11.9 μm
29.	e(P)	A	03 21 46	Near Coast of Venezuela
				10.54 N 63.42 W
				H = 03 10 17.0 h = normal MB=5.1 MS=5.3
				D = 72.43 Az = 40.2 (NEIS)
29.	eP diff	AB	03 28 44	Banda Sea 6.88 S 129.46 E
	ePKIKP	A	32 39	H = 03 14 14.6 h = 117 km MB = 6.5
	ePP	C	33 32	D = 112.77 Az = 322.4 (NEIS)
	e	C	34 30	PPV B 15s 9.0 μm M = 6.9
	e	C	38 50	LmH B 22 22.0 μm
	eSKKS	C	40 12	LmV B 18 12.0 μm
	e	C	42 15	
	eiSS	C	49 00	
	eSSS	C	52 25	
	LmH	B	04 17.9	
	LmV	B	21.5	
29.	ePKP	A	09 07 29	Samoa Islands Region 15.97 S 172.62 W
				H = 08 47 53.7 h = normal MB=4.8 MS=4.8
				D = 145.24 Az = 355.3 (NEIS)
29.	ePKHKP	A	10 11 03	South of Australia 51.82 S 139.44 E
	eX	A	11 17	H = 09 51 19.9 h = normal MB=5.0 MS=5.2
				D = 147.88 Az = 288.9 (NEIS)
				XV A 2.6s 190.6nm

Day	Phase		h m s	Remarks
30.	LmH	B	11 19.9	Hokkaido Region 44.57 N 143.30 E
	LmV	B	26.5	H = 10 33 51.1 h = 33 km MB = 4.8 (ISC)
				D = 76.2
				LmH B 15.5s 0.8 µm M = 5.1
				LmV B 14 0.4 µm 5.0
30.	eP	AC	16 20 01	Ryukyu Islands 29.85 N 130.43 E
	ePP	C	23 12	H = 16 07 33.2 h = 33 km MB=5.3 MS=5.8
	eS	C	30 30	D = 83.38 Az = 325.8 (NEIS)
	eSS	C	36 06	PV A 1.2s 32.5nm M = 5.3
	LmH	B	17 01.2	LmH B 17 13.0 µm 6.4
	LmV	B	01.2	LmV B 17 16.3 µm 6.5
30.	eP	A	18 04 06.5	Bonin Islands Region 26.45 N 140.45 E
	e(PP)	A	07 50	H = 17 51 48.7 h = 422 km MB = 5.0
				D = 90.93 Az = 330.3 (NEIS)
				PV A 1.2s 12.2nm M = 4.7
30.	eP	A	21 57 58	Celebes Sea 3.48 N 122.22 E
				H = 21 45 15.2 h = 623 km MB = 5.2
				D = 100.21 Az = 322.7 (NEIS)
31.	eP	A	04 36 06	Tibet 31.24 N 85.33 E
				H = 04 26 23.5 h = normal MB = 4.6
				D = 56.57 Az = 312.9 (NEIS)
31.	ePKHKP	AB	07 06 26	Tonga Islands Region 22.37 S 174.78 W
	e	A	06 30	H = 06 46 35.2 h = normal MB=4.9 MS=5.3
	LmH	C	08 16.4	D = 151.33 Az = 351.5 (NEIS)
	LmV	C	22.7	PKHKPV A 1.7s 24.2nm
				LmH B 20 0.6 µm M = 5.3
				LmV B 18 0.8 µm 5.6
31.	eP	A	07 18 19	Pyrenees 43.25 N 0.88 W
	e	B	22 06	H = 07 15 41.4 h = normal MB = 3.8
	LmH	B	23.2	D = 11.29 Az = 44.7 (NEIS)
	LmV	B	23.2	LmH B 8.5s 0.4 µm M = 2.1
				LmV B 9 0.5 µm

Day	Phase		h m s	Remarks
31.	e	A	08 36 46	
	e	A	38 07.5	
31.	eP	A	18 11 39	Fox Islands, Aleutian Is. 52.77 N 167.01 W H = 17 59 48.6 h = 36 km MB = 4.9 D = 76.95 Az = 0.9 (NEIS) PV A 1.3s 19.7nm M = 5.0
31.	ePn	A	22 25 21	Yugoslavia 44.64 N 18.35 E
	eSn	A	26 36	H = 22 23 22.7 h = 33 km MB = 4.0 (NEIS)
	eSg	A	27 32	D = 7.53

Day	Phase		h m s	Remarks
1.	ePKHKP	A	03 56 10	Tonga Islands 21.59 S 174.31 W H = 03 36 19.6 h = normal MB=4.7 MS=4.9 D = 150.63 Az = 352.3 (NEIS) PKHKPV A 1.6s 33.0nm
1.	ePg	A	08 02 53.5	D c. 175
	eSg	A	03 16	PgV A traces
1.	e(P)	A	10 42 58.5	
	e(Sg)	A	44 51	
1.	eP	A	12 55 18	Eastern Gulf of Aden 14.70 N 52.16 E H = 12 46 35.7 h = normal MB = 4.8 D = 48.50 Az = 326.5 (NEIS) PV A 1.6s 16.5nm M = 4.8
2.	eP	A	01 13 33	North Atlantic Ridge 10.34 N 40.92 W
	LmH	B	39.8	H = 01 03 36.7 h = 33 km MB = 4.9 (NEIS)
	LmV	B	40.0	D = 58.76 PV A 1.6s 33.0nm M = 5.2 LmH B 17 0.3 µm 4.5 LmV B 18 0.6 µm 4.9
2.	eP	A	03 11 14	Near East Coast of Kamchatka 53.94 N 160.36 E H = 02 59 51.2 h = 48 km MB = 4.6 D = 72.53 Az = 339.7 (NEIS) PV A 1.3s 13.1nm M = 4.7
2.	+iP	AB	05 05 50	Novaya Zemlya 70.82 N 54.06 E
	iPn	B	06 07	H = 04 59 56.7 h = 0 km MB=6.7 MS=5.3
	ePcP	B	08 56	D = 27.98 Az = 246.3 (NEIS)
	eS	B	10 35	PV A 1.2s 1972.0nm M = 6.8
	eSn	B	11 00	PV B 2.5 3.6 µm 6.8
	LmH	B	17.7	LmH B 11 23.2 µm 6.0
	LmV	B	17.7	LmV B 12 22.8 µm 6.1
	e	A	38 20	
	e(P'P')	A	39 26.5	

Day	Phase		h m s	Remarks
2.	eP	A	08 39 25	Hokkaido, Japan Region 41.43 N 142.10 E H = 08 27 29.2 h = 67 km MB = 5.1 D = 78.61 Az = 330.4 (NEIS) PV A 1.0s 15.8nm M = 4.9
2.	eP	A	12 51 19	Off East Coast of Honshu, Japan 39.55 N 143.42 E H = 12 39 07.6 h = 32 km MB = 4.7 D = 80.74 Az = 331.3 (NEIS)
2.	eP	A	19 55 29	Hokkaido, Japan Region 42.99 N 144.43 E H = 19 43 41.7 h = 121 km MB = 4.9 D = 78.09 Az = 331.5 (NEIS) PV A traces
2.	ePP	A	22 01 38.5	South of Java 10.10 S 112.64 E H = 21 43 16.8 h = 39 km MB = 5.4 (NEIS) D = 104.73
2.	eP epP	A A	22 07 41.5 07 55.5	Near East Coast of Honshu, Japan 36.34 N 141.70 E H = 21 55 20.4 h = 44 km MB = 5.1 D = 82.88 Az = 330.6 (NEIS) h = 52 km PV A 1.8s 27.0nm M = 5.0
2.	ePKP ePP LmH LmV	AB A B B	22 38 30.5 41 52 49.9 50.0	Tonga Islands 15.18 S 174.05 W H = 22 19 05.2 h = 97 km MB = 5.6 D = 144.33 Az = 353.8 (NEIS) PKPV A 1.5s 196.0nm PPV A 1.9 87.1nm M = 5.7 LmH B 14 0.8 µm LmV B 14 1.0 µm
2.	eP	A	22 57 32.5	Near East Coast of Honshu, Japan 36.43 N 141.88 E H = 22 45 07.9 h = 16 km MB = 4.9 D = 82.87 Az = 330.7 (NEIS)

Day	Phase		h m s	Remarks
3.	+eP	A	10 35 58	<u>Northern Sinkiang Prov., China</u>
	e	A	36 08.5	43.63 N 81.87 E
				H = 10 27 31.0 h = normal MB = 5.2
				D = 46.65 Az = 304.5 (NEIS)
				PV A 1.1s 20.2nm M = 5.0
3.	LmH	C	17 28.4	<u>Philippine Islands Region</u>
	LmV	C	29.5	15.04 N 122.61 E
				H = 16 29 13.0 h = 71 km MB = 4.7 (ISC)
				D = 91.3
				LmH C 17s 0.3/um
				LmV C 17 0.4/um
3.	e(PKIKP)	A	18 55 07	<u>Easter Island Cordillera</u>
				34.83 S 108.30 W
				H = 18 35 59.0 h = normal MB = 5.1
				D = 134.40 Az = 50.6 (NEIS)
3.	ePn	A	23 00 50	<u>Northern Italy</u> 44.39 N 12.22 E
	ePg	A	01 22	H = 22 59 18.3 h = 22 km
	eSn	A	01 58	D = 6.27 Az = 356.5 (NEIS)
	eSg	A	02 45	LmH C 20s 0.35/um M = 2.9
	LmH	C	18.0	LmV C 22 0.45/um
	LmV	C	18.3	
4.	ePKHKP	A	15 06 52	<u>Tonga Islands Region</u> 22.37 S 174.86 W
	ePKP2	A	07 03.5	H = 14 47 00.1 h = normal MB=5.0 MS=4.4
				D = 151.32 Az = 351.4 (NEIS)
				PKHKPV A 2.0s 51.2nm
4.	e(PKHKP)	A	17 49 28	<u>Hawaiian Islands Region</u>
	e(PKP2)	A	49 39.5	22.27 S 174.82 W
	LmV	C	19 04.5	H = 17 29 24.4 h = 33 km MB = 4.9 (NEIS)
				D = 151.23
5.	ePKP2	A	10 59 31	<u>Cook Strait, New Zealand</u>
	e	A	11 00 10.5	39.68 S 173.83 E
	LmH	C	12 10.0	H = 10 38 41.2 h = 33 km MB = 5.3 (NEIS)

Day	Phase		h m s	Remarks
cont. 5.	LmV	C	12 10.5	D = 163.39 LmH C 25s 0.6 /um M = 5.2 LmV C 24 0.7 /um 5.4
5.	ePKP2	A	19 41 46	West of Macquarie Islands 53.12 S 140.21 E H = 19 21 55.6 h = normal MB = 5.1 (NEIS) D = 148.73
6.	ePn	A	01 00 20	Federal Republic of Germany
	eSn	A	00 49	49.88 N 8.01 E
	eSg	A	01 00	H = 00 59 37.5 h = 0 km D = 2.44 Az = 70 (ISC)
6.	iPn	A	12 29 02.5	Federal Republic of Germany
	ePg	A	29 06.5	49.83 N 7.88 E
	eSn	A	29 32	H = 12 28 20.1 h = 8 km
	eSg	A	29 39.5	D = 2.53 Az = 69.8 (NEIS)
7.	eP	A	00 14 13	Kurile Islands 43.98 N 148.49 E H = 00 02 12.6 h = 29 km MB = 4.9 D = 78.57 Az = 333.6 (NEIS) PV A 1.0s 21.7nm M = 5.1
7.	ePn	A	02 35 41.5	Poland 50.17 N 18.45 E
	ePg	A	35 56.5	H = 02 34 32.2 h = 0 km
	eSg	A	36 55	D = 4.39 Az = 279 (ISC)
7.	e(P)	A	04 26 56	Southwest of Sumatra 0.94 S 97.22 E H = 04 13 56.3 h = 11 km MB = 4.9 D = 87.92 Az = 320.6 (NEIS)
7.	eP	A	05 11 04.5	Kurile Islands 43.32 N 156.27 E H = 04 59 07.9 h = 55 km MB = 5.0 D = 78.43 Az = 332.5 (NEIS)

Day	Phase		h m s	Remarks
7.	eP	A	13 30 53	<u>Southern Peru</u> 15.41 S 70.52 W
	LmH	C	14 10.0	H = 13 17 37.3 h = 153 km MB = 5.4
	LmV	C	10.0	D = 96.87 Az = 39.4 (NEIS)
				PV A 2.0s 29.9nm M = 5.4
				LmH C 20 0.4 µm
				LmV C 19 0.45 µm
7.	e(P)	A	20 14 16	<u>Komandorsky Islands Region</u>
				55.68 N 164.38 E
				H = 20 02 51.3 h = 42 km MB = 4.7
				D = 71.65 Az = 342.1 (NEIS)
8.	ePKP2	A	01 54 48	<u>West of Macquarie Island</u>
				59.25 S 149.60 E
				H = 01 34 31.0 h = normal
				D = 154.77 Az = 268.1 (NEIS)
8.	ePKP	A	13 54 12	<u>Tonga Islands</u> 15.57 S 173.22 W
	e	A	54 17	H = 13 34 35.3 h = 12 km MB=5.3 MS=4.9
	e	A	54 23.5	D = 144.80 Az = 354.7 (NEIS)
	ePP	A	57 46.5	PKPV A 1.5s 17.6nm
	e	A	57 50.5	LmH C 20 0.4 µm M = 5.1
	e	A	57 57.5	LmV C 20 0.4 µm 5.2
	LmH	C	15 01.0	
	LmV	C	05.8	
8.	+eiP	AB	21 35 04	<u>Hokkaido, Japan Region</u>
	epP	AB	35 36	42.49 N 141.78 E
	ePP	B	38 04	H = 21 23 21.8 h = 132 km MB = 6.0
	iS	B	44 44	D = 77.57 Az = 330.1 (NEIS)
	ePS	C	45 43	h = 131 km
	eSS	C	49.5	PV A 1.3s 528.0nm M = 6.1
	LmH	B	22 08.7	SH B 11.8 6.7 µm 6.5
	LmV	B	16.2	LmH B 15.5 10.6 µm
				LmV B 13.5 6.8 µm

Day	Phase		h m s	Remarks
9.	ePKP	A	01 49 30	Loyalty Islands Region 22.43 S 170.48 E
				H = 01 29 51 h = 31 km
				D = 147.36 Az = 335 (ISC)
				PKPV A 1.1s 18.1nm
9.	eP	A	05 57 57	Kurile Islands 48.12 N 152.88 E
				H = 05 46 24.2 h = 153 km MB = 4.8
				D = 76.14 Az = 335.8 (NEIS)
				PV A 1.2s 24.4nm M = 4.8
9.	eP	A	06 03 52.5	Greece-Albania Border Region
				38.96 N 20.26 E
				H = 06 00 44.9 h = 23 km MB = 4.8
				D = 13.19 Az = 335.2 (NEIS)
9.	eP	C	10 42 05	Near North Coast of Colombia
	eS	B	52 00	11.5 N 75.2 W
	eSS	B	56 56	H = 10 29 59.4 h = 0 km MB = 4.6
	LmH	B	11 13.7	D = 79.19 Az = 40 (ISC)
	LmV	B	13.7	PV C 9s 0.6/um M = 5.6
				LmH B 19.5 11.3/um 6.2
				LmV B 21 12.7/um 6.3
9.	eP	A	10 47 32	Near East Coast of Kamchatka
				53.29 N 160.29 E
				H = 10 36 03.6 h = normal MB=4.7 MS=5.3
				D = 73.13 Az = 339.8 (NEIS)
9.	eP	A	12 18 38	Algeria 36.38 N 5.30 E
				H = 12 15 07.0 h = normal MB = 4.1
				D = 14.97 Az = 15.7 (NEIS)
9.	eP	AB	13 13 34	Near Coast of Peru 12.50 S 77.79 W
	ePP	AB	17 33	H = 12 59 49.8 h = 6 km MB=6.0 MS=7.2
	eSKS	B	24 14	D = 99.17 Az = 40.2 (NEIS)
	eS	B	25 04	PV A 1.8s 81.0nm M = 6.1
	ePS	B	26 40	PV B 14 7.0/um 7.2
	eSS	B	31 40	PPV A 4.5 1397.0nm 6.7

Day	Phase		h m s	Remarks
cont. 9.	LmV	B	13 55.3	PPV B 16s 12.2/um M = 8.0
	LmH	B	56.2	LmH B 19.8 90.0/um 7.3
				LmV B 17 93.6/um 7.4
9.	eP	A	14 28 47	<u>Off East Coast of Honshu, Japan</u>
				40.02 N 143.35 E
				H = 14 16 36.8 h = 27 km MB = 5.4
				D = 80.31 Az = 331.2 (NEIS)
				PV A 1.6s 60.5nm M = 5.4
9.	eP	AB	19 24 26	<u>Sunda Strait</u> 6.50 S 105.34 E
	ePP	AB	28 21	H = 19 10 55.2 h = 51 km MB = 6.1
	eSKS	B	35 10	D = 97.34 Az = 320.2 (NEIS)
	eS	B	35 44	PV A 2.0s 85.5nm M = 5.9
	eSS	C	42 32	PPV A 1.9 303.0nm 6.4
	LmH	B	20 16.5	LmH B 17 4.9/um 6.1
	LmV	B	16.5	LmV B 17 3.9/um 6.0
9.	e(Sg)	A	20 21 09.5	
9.	ePKP	A	23 07 46	<u>Fiji Islands Region</u> 17.56 S 178.49 W
				H = 22 49 10.5 h = 590 km MB = 4.8
				D = 146.01 Az = 348.5 (NEIS)
				PKPV A 1.4s 14.0nm
10.	ePKP	AB	04 45 09	<u>Fiji Islands Region</u> 15.86 S 178.51 W
	LmH	B	05 47.8	H = 04 25 31.9 h = 33 km MB=5.8 MS=6.1
	LmV	B	53.6	D = 144.35 Az = 348.9 (NEIS)
				PKPV A 1.6s 44.0nm
				LmH B 19 3.7/um M = 6.1
				LmV B 22 3.0/um 6.0
11.	ePn	A	00 41 51	<u>Fed. Republic of Germany</u> 48.2 N 8.7 E
	ePg	A	42 00	H = 00 41 09 h = 0 km
	eSg	A	42 36	D = 3.05 Az = 37 (ISC)

Day	Phase		h m s	Remarks
11.	ePKP	A	04 30 55	Fiji Islands Region 17.87 S 178.47 W
				H = 04 12 18.7 h = 593 km MB = 5.2
				D = 146.32 Az = 348.4 (NEIS)
				PKPV A 1.6s 27.5nm
				51.63 N
11.	eP	A	05 29 41.5	Andreanof Islands, Aleutian Is. 178.11 W
	eS	C	39 47	H = 05 17 51.0 h = 68 km MB = 5.8
	eSS	C	44 56	D = 77.77 Az = 353.7 (NEIS)
	LmV	C	59.0	PV A 1.5s 70.4nm M = 5.4
	LmH	C	06 00.0	LmH C 25 1.3/um 5.1
				LmV C 26 1.6/um 5.3
11.	ePKIKP	A	06 48 47	South of Fiji Islands 23.91 S 177.56 W
	-eiPKHKP	A	48 55	H = 06 29 21.1 h = 196 km MB = 5.6
	+iPKP2	A	49 06	D = 152.37 Az = 347.4 (NEIS)
				PKIKPV A 1.4s 34.9nm
				PKHKPV A 1.3 76.4nm
				PKP2V A 1.4 121.0nm
11.	LmH	C	21 07.6	Santa Cruz Islands 10.64 S 166.38 E
	LmV	C	17.6	H = 19 53 15 h = 51 km MB = 4.6 (ISC)
				D = 135.0
				LmH C 24s 0.45/um M = 5.1
				LmV C 18 0.45/um 5.2
12.	LmH	C	01 52.8	Off Coast of Peru 10.75 S 79.11 W
	LmV	C	57.8	H = 00 53 55 h = 45 km MB = 4.9 (ISC)
				D = 98.7
				LmH C 18s 0.4/um M = 5.0
				LmV C 16 0.7/um 5.3
12.	ePn	A	02 59 39	France 48.34 N 6.77 E
	iPg	A	59 52.5	H = 02 58 40.1 h = 36 km
	iSg	A	00 46	D = 3.91 Az = 52.0 (NEIS)
12.	eP	A	03 28 28	Kurile Islands 45.46 N 150.91 E
	LmH	C	04 06.0	H = 03 16 31.7 h = normal MB=4.8 MS=4.3
	LmV	C	06.0	D = 77.99 Az = 334.9 (NEIS)
				PV A 1.4s 23.3nm M = 5.0
				LmH C 18 0.45/um 4.8
				LmV C 19 0.6/um 5.0

Day	Phase		h m s	Remarks
12.	ePKP	A	03 50 03.5	New Hebrides Islands 20.14 S 169.26 E
				H = 03 30 32.5 h = 46 km MB = 4.8
				D = 144.80 Az = 335.2 (NEIS)
				PV A 1.2s 20.3nm
12.	eP	A	22 27 09	Celebes Sea 2.32 N 121.13 E
	epP	A	27 27	H = 22 13 25.8 h = 54 km MB = 5.8
	ePP	A	31 28	D = 100.46 Az = 322.4 (NEIS)
	e	A	31 40	h = 65 km
	ePKKP	A	43 43.5	PV A 1.6s 35.8nm M = 5.7
	LmH	B	23 19.0	PKKPV A 1.4 16.3nm
	LmV	B	20.6	LmH B 19 1.2 μm
				LmV B 18 1.1 μm
13.	eP	A	02 41 46.5	Eastern Caucasus 42.72 N 46.56 E
	LmV	C	51.9	H = 02 36 25.5 h = 42 km MB=5.1 MS=4.7
	LmH	B	52.5	D = 25.06 Az = 300.6 (NEIS)
				PV A 1.4s 79.0nm M = 5.1
				LmH B 14 2.0 μm 4.8
				LmV C 18 1.0 μm 4.5
13.	ePKP2	A	05 43 30.5	South of Fiji Islands 25.77 S 176.31 W
				H = 05 23 16.7 h = normal MB=4.9 MS=4.4
				D = 154.43 Az = 348.3 (NEIS)
13.	ePKHKP	A	17 19 15	West of Macquarie Island
	ePKP2	A	19 28.5	57.96 S 148.28 E
	LmH	B	18 37.0	H = 16 59 16.6 h = normal MB=5.3 MS=5.9
	LmV	B	42.3	D = 154.07 Az = 272.5 (NEIS)
				PKHKPV A 1.6s 27.5nm
				LmH B 19 1.2 μm M = 5.6
				LmV B 22 1.8 μm 5.8
14.	eP1	A	05 00 05	Alaska Peninsula 58.80 N 154.62 W
	eP2	A	00 22	H = 04 48 54.7 h = 37 km MB=5.5 MS=5.6
	eP3	A	00 46	D = 70.34 Az = 9.3 (NEIS)
	LmH	B	35.6	P1V A 1.2s 16.3nm M = 4.9

Day	Phase		h m s	Remarks
cont. 14.	LmV	B	05 35.6	P2V A 1.4s 41.9nm M = 5.3
				P3V A 1.7 66.7nm 5.4
				LmH B 17 1.7/um 5.4
				LmV B 18 2.0/um 5.5
14.	eP	A	13 26 02	<u>Greece</u> 38.52 N 23.12 E
	eS	B	29 04	H = 13 22 33.1 h = 19 km MB=5.0 MS=5.0
	LmH	B	31.0	D = 14.61 Az = 329.8 (NEIS)
	LmV	B	33.2	PV A 1.5s 45.2nm M = 4.8
				LmH B 12 17.0/um 5.3
				LmV B 12 10.8/um
14.	eP	A	14 30 15	<u>Greece</u> 38.53 N 22.99 E
	eS	C	33.3	H = 14 26 45.8 h = 3 km MB = 5.2
	LmH	B	35.3	D = 14.55 Az = 330.0 (NEIS)
	LmV	B	37.4	LmH B 13s 11.2/um M = 5.2
				LmV B 14 5.3/um
14.	eP	A	15 33 13	<u>Greece</u> 38.49 N 23.09 E
	LmH	B	38.2	H = 15 29 44.8 h = 24 km MB=5.0 MS=5.1
	LmV	B	40.5	D = 14.63 Az = 329.9 (NEIS)
				PV A 1.4s 18.6nm M = 4.4
				LmH B 12.5 7.8/um 5.0
				LmV B 13 4.4/um
14.	ePn	A	17 10 10	<u>Svabian Jura Region, Fed. Rep. of Germany</u>
	ePg	A	10 17	48.31 N 9.12 E
	eSg	A	10 55	H = 17 09 24.3 h = 15 km
				D = 2.85 Az = 34.0 (NEIS)
15.	eP	A	00 26 45	<u>Eastern Gulf of Aden</u> 13.49 N 50.21 E
				H = 00 18 02.5 h = normal MB = 4.5
				D = 48.49 Az = 328.0 (NEIS)
				PV A traces
15.	LmH	C	16 15.2	
	LmV	C	15.3	

Day	Phase		h m s	Remarks
15.	eP	A	17 45 48	Southern Iran 27.63 N 62.45 E
	e	A	46 14	H = 17 37 45.8 h = 75 km MB = 4.7
				D = 44.57 Az = 315.3 (NEIS)
				PV A traces
15.	eP	A	22 40 01	Southern Sinkiang Prov., China
				37.40 N 76.91 E
				H = 22 31 29.2 h = 34 km MB = 5.0
				D = 47.29 Az = 308.1 (NEIS)
15.	+iP	AB	23 45 05.5	Near East Coast of Honshu, Japan
	ePP	A	48 12	35.77 N 140.99 E
	eS	B	55 20	H = 23 32 42.1 h = 36 km MB=5.8 MS=5.6
	eSS	C	24 00 52	D = 83.09 Az = 330.3 (NEIS)
	LmH	B	26.9	PV A 1.0s 157.0nm M = 6.1
	LmV	B	27.3	PPV A 1.3 54.6nm 5.8
				LmH B 16 3.4 /um 5.8
				LmV B 14 3.1 /um 5.9
16.	eP	A	10 51 18	Molucca Passage 2.43 N 127.71 E
	e	A	54 40	H = 10 37 08.0 h = 36 km MB=5.6 MS=5.4
	ePP	AB	55 32	D = 104.31 Az = 323.8 (NEIS)
	LmH	B	11 34.2	PV A 1.2s 24.4nm M = 5.9
	LmV	B	44.1	PPV A 1.5 40.2nm 5.8
				LmH B 23 1.4 /um 5.5
				LmV B 18 0.6 /um 5.2
16.	eP	A	16 27 24	Kashmir-India Border Region
				32.85 N 76.14 E
				H = 16 18 36.6 h = 63 km MB = 4.8
				D = 49.70 Az = 311.1 (NEIS)
16.	eP1	A	16 36 43	Kansu Province, China 33.05 N 103.98 E
	eP2	A	36 48.5	H = 16 25 53.8 h = normal MB=5.1 MS=5.2
	LmH	B	17 02.6	D = 66.64 Az = 316.1 (NEIS)
	LmV	B	07.4	P2V A 1.8s 40.6nm M = 5.2
				LmH B 17 2.0 /um 5.4
				LmV B 15 1.3 /um 5.3

Day	Phase		h m s	Remarks
16.	eP	A	19 29 54	North Atlantic Ocean 52.65 N 32.12 W
	eS	C	34 30	H = 19 24 14.5 h = normal MB=5.0 MS=4.9
	LmH	B	40.1	D = 26.90 Az = 76.6 (NEIS)
	LmV	B	40.7	PV A 1.0s 23.6nm M = 4.8
				LmH B 15 4.0 μm 5.1
				LmV B 14 3.2 μm 5.2
16.	eP	A	22 18 05	Burma 24.88 N 95.42 E
				H = 22 07 25.1 h = 159 km MB = 4.7
				D = 67.36 Az = 316.7 (NEIS)
17.	eP	A	00 17 30	Panama-Colombia Border Region
				7.74 N 77.60 W
				H = 00 05 00.4 h = 21 km MB=5.0 MS=4.3
				D = 83.58 Az = 39.8 (NEIS)
17.	+eiPKP	A	01 19 55	Tonga Islands 16.99 S 174.26 W
	epPKP	A	20 45	H = 01 00 36.5 h = 192 km MB = 5.0
				D = 146.09 Az = 353.3 (NEIS)
				h = 193 km
				PKPV A 1.7s 103.0nm
17.	eP	A	07 45 29	North Atlantic Ocean 52.72 N 32.00 W
				H = 07 39 50.9 h = normal MB = 4.5
				D = 26.82 Az = 76.8 (NEIS)
17.	eP	A	15 12 49	Iran 32.81 N 55.07 E
	LmH	C	31.2	H = 15 05 47.7 h = 43 km MB = 5.2
				D = 36.39 Az = 312.4 (NEIS)
				LmH C 24s 0.5 μm M = 4.2
17.	e	A	15 40 12	Off Coast of Oregon 43.50 N 127.04 W
				H = 15 27 59.4 h = 12 km MB = 5.1
				D = 79.57 Az = 25.3 (NEIS)
17.	+iP	AB	17 35 40	Near East Coast of Kamchatka
				54.81 N 161.54 E
				H = 17 24 17.9 h = 31 km MB = 5.3
				D = 71.95 Az = 340.4 (NEIS)
				PV A 0.9s 85.6nm M = 5.8

Day	Phase		h m s	Remarks
18.	eP	A	07 27 23	<u>Near East Coast of Kamchatka</u> 55.31 N 161.87 E H = 07 16 04.0 h = normal MB = 4.6 D = 71.54 Az = 340.5 (NEIS)
18.	eP	A	18 15 41	<u>Philippine Islands Region</u>
	LmH	B	59.5	20.57 N 121.19 E
	LmV	B	59.7	H = 18 03 03.3 h = 36 km MB = 5.5 D = 86.02 Az = 323.0 (NEIS) PV A 1.6s 49.5nm M = 5.5 LmH B 18 1.1 µm 5.3 LmV B 16 0.9 µm 5.3
19.	+eP	AB	04 08 02	<u>Philippine Islands Region</u>
	epP	A	08 13.5	18.97 N 121.31 E
	ePP	C	11 30	H = 03 55 18.9 h = 44 km MB =5.7 MS=5.7
	eSKS	C	18 24	D = 87.36 Az = 323.1 (NEIS)
	eS	C	18 36	PV A 2.4s 165.8nm M = 5.8
	eSS	C	24 28	LmH B 18 2.9 µm 5.7
	LmH	B	52.7	LmV B 19 4.6 µm 5.9
	LmV	B	52.7	
19.	ePP	AC	05 18 53	<u>New Ireland Region</u> 3.19 S 150.64 E
	ePS	B	28 48	H = 04 58 23.0 h = 18 km MB=5.5 (NEIS)
	eSS	C	35 30	D = 121.5
	LmH	B	06 12.7	PPV A 2.6s 139.0nm M = 6.1
	LmV	B	13.0	LmH B 20 6.6 µm 6.3 LmV B 20 6.2 µm 6.3
19.	ePKIKP	A	05 59 00	<u>Fiji Islands Region</u>
	ePKHKP	A	59 02.5	17.95 S 178.69 W H = 05 40 29.6 h = 639 km MB = 5.1 D = 146.35 Az = 348.1 (NEIS) PKHKPV A 1.2s 24.4nm
19.	ePKIKP	A	07 09 54	<u>New Hebrides Islands</u> 16.07 S 167.46 E H = 06 50 29.0 h = 42 km MB = 5.2 D = 140.41 Az = 335.9 (NEIS) PKIKPV A 1.4s 18.6nm

Day	Phase		h m s	Remarks
20.	eP	A	00 20 58	<u>Fox Islands, Aleutian Is.</u>
	epP	A	21 11.5	53.60 N 165.25 W
				H = 00 09 15.0 h = 57 km MB = 5.0
				D = 76.10 Az = 2.1 (NEIS)
				h̄ = 50 km
20.	eP	A	00 51 44.5	<u>Northeast of Taiwan</u> 25.13 N 123.47 E
				H = 00 39 33.5 h = 173 km MB = 4.9
				D = 83.63 Az = 323.5 (NEIS)
20.	ePKIKP	AB	04 34 04	<u>New Hebrides Islands</u> 15.02 S 167.14 E
	ePKIKPm	AB	34 34	H = 04 14 46.9 h = normal MB=6.4 MS=6.9
	eSS	B	55 50	D = 139.33 Az = 336.1 (NEIS)
	e	B	57 45	PKIKPV A 2.0s 598.3nm
	LmH	B	05 37.8	PKIKPV B 13 8.1/um
	LmV	B	44.2	LmH B 23 45.2/um M = 7.1
				LmV B 19 29.8/um 7.0
20.	ePn	A	08 43 05.5	<u>Austria</u> 46.99 N 11.31 E
	ePg	A	43 25.5	H = 08 42 07.4 h = normal
	eSn	A	43 56.5	D = 3.66 Az = 3.0 (NEIS)
	eSg	A	44 24.5	
20.	ePP	A	13 40 44	<u>Southwestern Atlantic Ocean</u>
	LmH	B	14 14.7	53.57 S 28.35 W
	LmV	B	24.6	H = 13 21 41.2 h = 33 km MB = 6.0 (NEIS)
				D = 109.14
				PPV A 1.6s 22.0nm M = 5.6
				LmH B 19 1.0/um 5.4
				LmV B 18 0.6/um 5.2
20.	ePn	A	15 12 06.5	<u>Northern Italy</u> 45.77 N 12.26 E
	ePg	A	12 22	H = 15 10 54.9 h = normal
	eSn	A	13 00	D = 4.90 Az = 355.2 (NEIS)
	eiSg	A	13 29.5	

Day	Phase		h m s	Remarks
20.	ePn	A	15 12 06.5	Northern Italy 45.77 N 12.26 E
	ePg	A	12 22	H = 15 10 54.9 h = normal
	eSn	A	13 00	D = 4.90 Az = 355.2 (NEIS)
	eiSg	A	13 29.5	
20.	ePn	A	19 34 15	Northern Italy 45.77 N 12.19 E
	ePg	A	34 36	H = 19 33 02.4 h = 25 km
	eSn	A	35 08	D = 4.90 Az = 355.7 (NEIS)
	eSg	A	35 37	
21.	LmH	C	08 42.9	Solomon Islands 7.99 S 155.67 E
	LmV	C	51.2	H = 07 32 48 h = 19 km MB = 5.1 (ISC)
				D = 128.2
				LmH C 24s 0.5 μm M = 5.1
				LmV C 24 0.4 μm 5.0
21.	ePKP	A	13 10 00	Tonga Islands 16.75 S 174.37 W
	epPKP	A	10 50	H = 12 50 38.2 h = 160 km MB = 4.7
				D = 145.85 Az = 353.2 (NEIS)
				h = 190 km
				PKPV A 1.4s 18.6nm
21.	eP	A	19 56 20.5	Burma 20.07 N 95.18 E
	e	A	56 41.5	H = 19 45 05.6 h = normal MB = 5.1
				D = 70.72 Az = 317.9 (NEIS)
21.	eP	A	21 48 56.5	North Atlantic Ridge 52.27 N 31.61 W
	LmH	B	58.9	H = 21 43 19.6 h = normal MB = 4.7
	LmV	B	59.6	D = 26.69 Az = 76.1 (NEIS)
				LmH B 15s 0.6 μm M = 4.3
				LmV B 14 0.7 μm 4.5
21.	e	A	22 07 43.5	
22.	eSg	A	10 31 45	
22.	ePn	A	17 04 29.5	Northern Italy 46.00 N 12.16 E
	ePg	A	04 48.5	H = 17 03 18.6 h = 0 km

Day	Phase		h m s	Remarks
cont. 22.	eSn	A	17 05 22	D = 4.66 Az = 356 (ISC)
	eSg	A	05 47.5	
23.	ePKIKP	A	00 55 32.5	South of Fiji Islands 22.79 S 177.54 W
	-iPKHKP	AB	55 38.5	H = 00 36 08.9 h = 211 km MB = 5.4
	-iPKP2	AB	55 42	D = 151.29 Az = 347.8 (NEIS)
				PKIKPV A 1.2s 16.3nm
				PKHKPV A 1.0 126.0nm
				PKP2V A 1.2 89.4nm
23.	eP	A	03 38 37	Iran-USSR Border Region 38.2 N 55.3 E
				H = 03 31 58 h = 0 km MB = 4.1
				D = 33.12 Az = 306 (ISC)
23.	ePn	A	07 55 28	Southern Italy 39.93 N 18.91 E
	ePg	A	56 26	H = 07 52 28.9 h = normal MB = 4.4
	eSn	A	57 31	D = 11.88 Az = 336.9 (NEIS)
23.	eP	A	09 56 38	Southwestern Ryukyu Islands
	e	A	56 48.5	23.49 N 123.75 E
	LmH	B	10 39.7	H = 09 44 03.8 h = normal MB = 5.4 MS=5.1
	LmV	B	39.3	D = 85.10 Az = 323.7 (NEIS)
				PV A 1.6s 55.0nm M = 5.5
				LmH B 18.5 0.8/um 5.2
				LmV B 20 1.0/um 5.2
23.	eP	A	18 49 26.5	Greece - Albania Border Region
	e	A	50 05	39.71 N 19.08 E
	eS	C	52 30	H = 18 46 33.4 h = normal MB = 4.7
	LmH	B	53.6	D = 12.13 Az = 336.8 (NEIS)
	LmV	B	54.4	LmH B 15s 1.7/um M = 4.2
24.	LmH	C	07 50.5	South of Panama 5.58 N 82.61 W
	LmV	C	50.5	H = 07 05 37.2 h = 36 km MB=5.0 MS=4.9
				D = 88.41 Az = 39.4 (NEIS)
				LmH C 25s 0.9/um M = 5.1

Day	Phase		h m s	Remarks
25.	eP	A	16 44 31	Eastern Gulf of Aden 13.48 N 51.51 E H = 16 35 44.5 h = normal MB = 5.0 D = 49.17 Az = 327.3 (NEIS) PV A 1.6s 30.2nm M = 5.1
26.	esPKP	A	23 23 11.5	South of Fiji Islands 24.14 S 179.88 E H = 23 02 07.4 h = 516 km MB = 5.0 D = 152.01 Az = 344.0 (NEIS) sPKPV A 1.2s 12.2nm
27.	eP	A	07 45 35	Iran - Iraq Border Region 33.21 N 46.90 E H = 07 39 15.8 h = 31 km MB = 4.9 D = 31.13 Az = 314.6 (NEIS)
27.	ePn	A	12 57 04.5	Austria 47.05 N 10.45 E
	ePg	A	57 18.5	H = 12 56 09.6 h = normal
	eSn	A	57 47.5	D = 3.69 Az = 11.7 (NEIS)
	eSg	A	58 04.5	
27.	eP	A	16 58 50	Iran - Iraq Border Region
	LmH	B	17 12.6	35.30 N 45.66 E
	LmV	B	13.6	H = 16 52 49.9 h = 50 km MB = 5.0 D = 28.94 Az = 312.6 (NEIS) LmH B 16s 1.1 µm M = 4.6 LmV B 15 1.0 µm 4.7
27.	eP	A	20 46 14.5	East of Severnaya Zemlya 79.25 N 123.99 E H = 20 38 03.4 h = normal MB = 4.4 D = 44.61 Az = 303.0 (NEIS)
28.	eP	A	01 59 43.5	Montana 47.69 N 113.01 W H = 01 48 23.8 h = 5 km D = 71.09 Az = 33.6 (NEIS)
28.	ePKP	A	05 27 33	Fiji Islands Region 17.88 S 178.69 W H = 05 08 51.4 h = 535 km MB = 4.6 D = 146.29 Az = 348.2 (NEIS)

Day	Phase		h m s	Remarks
28.	eP	A	05 40 40	Andreanof Islands, Aleutian Is. 51.87 N 175.27 W H = 05 28 48.3 h = 63 km MB = 5.2 D = 77.69 Az = 355.5 (NEIS) PV A 1.2s 14.2nm M = 4.8
28.	eP	A	13 40 16.5	Azores Islands Region 37.56 N 31.99 W H = 13 33 38.8 h = normal MB=4.8 MS=4.8 D = 33.44 Az = 52.8 (NEIS)
28.	ePKHKP ePKP2	A A	15 33 56.5 34 03.5	Fiji Islands Region 21.21 S 179.03 W H = 15 15 15.4 h 3 625 km MB = 4.8 D = 149.45 Az = 346.6 (NEIS) PKHKPV A 1.8s 47.3nm PKP2V A 1.0 15.7nm
28.	+eiP	A	16 43 44.5	Unimak Island Region 53.62 N 163.70 W H = 16 31 58.3 h = 32 km MB = 5.3 D = 76.03 Az = 3.1 (NEIS) PV A 1.2s 52.8nm M = 5.4
28.	LmH	B	23 49.1	Central California 36.91 N 121.50 W H = 23 01 24.8 h = 9 km MB=5.0 MS=4.5 D = 83.55 Az = 27.9 (NEIS) LmH B 18s 1.0 µm M = 5.2
29.	eP e	A A	00 54 18.5 54 27.5	Central Mid-Atlantic Ridge 3.62 N 31.86 W H = 00 44 18.4 h = normal MB = 4.7 D = 59.38 Az = 30.6 (NEIS) PV A 1.7s 24.2nm M = 5.1
29.	+ePKP epPKP	A A	10 08 34 09 11	New Hebrides Islands 19.56 S 169.36 E H = 09 49 15.4 h = 142 km MB = 5.5 D = 144.32 Az = 335.6 (NEIS) h = 140 km PKPV A 1.3s 61.1nm pPKPV A 1.6 30.2nm

Day	Phase		h m s	Remarks
29.	eP	A	21 14 36	USSR-Mongolia Border Region
				51.75 N 98.94 E
				H = 21 05 31.6 h = normal MB = 5.2
				D = 51.50 Az = 305.6 (NEIS)
				PV A 1.4s 34.9nm M = 5.1
29.	-iP	AB	22 17 19	South of Honshu, Japan 30.70 N 138.32 E
	ipP	AB	18 59	H = 22 05 22.4 h = 419 km MB = 6.1
	esP	B	19 36	D = 86.33 Az = 329.2 (NEIS)
	ePP	B	20 50	h = 432 km
	eS	B	27 12	PV A 1.5s 482.4nm M = 6.0
	esS	B	30 04	PV B 8 7.4/um 6.5
	eSS	B	33 12	LmH B 15 38.0/um
	eP'P	A	43 23	LmV B 15 41.6/um
	LmH	B	23 02.5	
	LmV	B	02.7	
30.	eP	A	13 09 03.5	Near Islands, Aleutian Is.
				53.27 N 172.96 E
				H = 12 57 20.6 h = 17 km MB=5.2 MS=4.9
				D = 75.28 Az = 347.8 (NEIS)
30.	LmV	B	15 15.6	Probably East of North Island, New Zealand
	LmH	B	15.7	35.42 S 179.4 W
				H = 13 55 01 h = 33 km MB = 5.0 (NEIS)
				D = 163.0
				LmH B 16s 0.7/um M = 5.5
				LmV B 16 1.1/um 5.8

Day	Phase		h m s	Remarks
1.	ePn	A	01 23 26	Central Italy 42.80 N 12.99 E
	e	A	24 51	H = 01 21 29.5 h = 11 km
	e	A	25 01	D = 7.90 Az = 354 (ISC)
1.	ePKIKP	A	02 11 58.5	Fiji Region 19.63 S 178.15 W
				H = 01 53 17.2 h = 580 km MB = 4.3
				D = 148.09 Az = 348 (ISC)
1.	eP	A	06 24 45	Greece 38.51 N 23.24 E
	e	A	24 52.5	H = 06 21 18.1 h = normal MB = 4.2
	LmH	B	29.7	D = 14.67 Az = 329.6 (NEIS)
				PV A traces
				LmH B 13s 0.9 µm M = 4.1
1.	eP	A	12 13 04	Turkey 39.52 N 26.20 E
	LmH	B	19.7	H = 12 09 28.8 h = 36 km MB = 4.5
	LmV	B	19.7	D = 15.13 Az = 322.1 (NEIS)
				PV A 2.6s 104.0nm M = 5.1
				LmH B 13 0.9 µm 4.1
				LmV B 12 1.5 µm 4.5
1.	eP	A	13 22 26.5	Mindanao, Philippine Islands
				5.99 N 123.83 E
				H = 13 09 40.1 h = 521 km MB = 5.1
				D = 99.18 Az = 323.3 (NEIS)
1.	ePKP2	A	21 58 35.5	South of Fiji 23.76 S 179.76 E
				H = 21 39 31 h = 547 km
				D = 151.63 Az = 344 (ISC)
1.	ePKP	A	22 45 49	Tonga Islands 16.22 S 173.36 W
				H = 22 26 12.5 h = normal MB = 4.5
				D = 145.43 Az = 354.4 (NEIS)
2.	ePKP2	A	00 38 39	East of North Island, New Zealand
				35.22 S 179.65 W
				H = 00 18 02.7 h = 120 km MB = 5.1 (NEIS)
				D = 162.55

Day	Phase		h m s	Remarks
2.	ePn	A	01 16 29	<u>Central Italy</u> 42.78 N 13.10 E
	eSn	A	17 54.5	H = 01 14 33.8 h = normal
	eSg	A	18 55	D = 7.93 Az = 353.1 (NEIS)
2.	ePn	A	01 57 11	<u>Central Italy</u> 42.99 N 12.97 E
	eSn	A	58 40	H = 01 55 15.9 h = 7 km MB = 4.9
	eSg	A	59 37	D = 7.72 Az = 353.6 (NEIS)
	LmH	B	02 00.9	PnV A 1.2s 24.4nm M = 5.3
	LmV	B	01.2	LmH B 14 1.6 µm 3.8
				LmV B 12 2.1 µm
2.	ePKIKP	A	02 10 39.5	<u>Santa Cruz Islands</u> 11.01 S 166.36 E
				H = 01 51 37.0 h = 153 km MB = 5.2
				D = 135.37 Az = 337.3 (NEIS)
2.	eP	A	06 46 49	<u>Philippine Islands Region</u>
	e	A	46 57	19.08 N 121.16 E
	eS	C	57 20	H = 06 34 07.7 h = 53 km MB = 5.5
	ePS	C	58 08	D = 87.18 Az = 323.1 (NEIS)
	LmH	B	07 29.5	PV A 1.5s 35.2nm M = 5.4
	LmV	B	29.5	LmH B 17 1.9 µm
				LmV B 16 2.2 µm
2.	ePKIKP	A	07 02 30	<u>New Britain Region</u> 6.17 S 153.07 E
	ePP	A	04 21	H = 06 43 30.4 h = 28 km MB = 5.8
				D = 125.27 Az = 330.9 (NEIS)
				PKIKPV A 1.6s 66.0nm
				PPV A 1.6 99.0nm, M = 6.0
2.	+eP	A	09 13 18.5	<u>Southern Iran</u> 27.99 N 55.82 E
	LmH	B	30.4	H = 09 05 44.2 h = 36 km MB = 5.4
	LmV	B	36.4	D = 40.21 Az = 316.6 (NEIS)
				PV A 1.4s 32.6nm M = 4.9
				LmH B 20 1.8 µm 4.9
				LmV B 14 1.0 µm 4.9
2.	ePKHKP	A	12 41 00	<u>Fiji Islands Region</u> 20.43 S 178.14 W
				H = 12 22 15.7 h = 587 km MB = 4.8
				D = 148.87 Az = 348.0 (NEIS)

Day	Phase	h m s	Remarks
2.	eP A	12 49 19	Guatemala 14.63 N 91.51 W
	epP A	49 45	H = 12 36 44.5 h = 97 km MB = 5.2
	eSKS C	59 40	D = 86.93 Az = 38.4 (NEIS)
	ePS C	13 01 04	h = 98 km
	LmH C	18.5	PV A 1.3s 30.6nm M = 5.2
	LmV C	18.5	LmV C 20 1.3/um
2.	ePn A	21 02 07.5	Central Italy 42.80 N 13.04 E
	eSn A	03 40	H = 21 00 11.3 h = 15 km
			D = 7.91 Az = 353 (ISC)
2.	eP A	23 05 44	Greece 38.48 N 22.19 E
	LmH B	11.7	H = 23 02 16.4 h = 16 km MB = 4.7
	LmV B	11.7	D = 14.29 Az = 331.8 (NEIS)
			LmH B 14s 1.1/um M = 4.1
			LmV B 10 1.2/um
3.	ePKP A	03 25 09	Banda Sea 5.01 S 129.78 E
	ePP A	26 02	H = 03 06 35.2 h = normal MB=6.2 MS=6.5
	LmH B	04 24.1	D = 111.49 Az = 322.9 (NEIS)
	LmV B	24.1	PPV A 1.6s 76.9nm M = 5.2
			LmH B 18 9.9/um 6.4
			LmV B 18 8.9/um 6.4
3.	ePn A	19 10 20	Central Italy 42.78 N 13.04 E
	eSn A	11 45	H = 19 08 12.7 h = 5 km
			D = 7.93 Az = 353 (ISC)
3.	eP A	19 30 37.5	Kurile Islands 48.08 N 153.75 E
			H = 19 18 50.2 h = 36 km MB = 5.0
			D = 76.41 Az = 336.3 (NEIS)
			PV A 1.0s 11.8nm M = 4.8
4.	e A	01 44 00	
4.	eP A	02 46 18	Ascension Island Region 7.74 S 13.52 W
	e A	46 23	H = 02 35 58.9 h = normal MB = 4.9
			D = 63.00 Az = 17.8 (NEIS)
			PV A 1.5s 32.7nm M = 5.2

Day	Phase		h m s	Remarks
4.	eP1	AB	03 20 33	Northern Sumatra 0.39 N 97.84 E
	eP2	AB	20 37	H = 03 07 46.3 h = 20 km MB=6.0 MS=6.9
	eS	B	31 08	D = 87.30 Az = 320.5 (NEIS)
	eSS	C	37 00	P1V A 1.6s 38.5nm M = 5.4
	LQ	C	46 20	P2V A 2.4 415.0nm 6.3
				P2V B 11 3.1 μm 6.5
				LmH B 19 21.4 μm 6.6
				LmV B 18 17.6 μm 6.5
4.	ePKHKP	A	05 30 02.5	Fiji Islands Region 18.34 S 177.06 W
				H = 05 11 02.4 h = 375 km MB = 4.7
				D = 147.03 Az = 349.8 (NEIS)
				PKHKPV A 1.8s 30.4nm
4.	eP	A	10 12 12	Kurile Islands 44.35 N 147.63 E
				H = 10 00 13.7 h = normal MB = 4.8
				D = 77.97 Az = 333.1 (NEIS)
4.	ePn	A	16 16 51	Central Italy 42.65 N 13.09 E
	eSn	A	18 29	H = 16 14 46.9 h = normal MB = 4.0
				D = 8.06 Az = 353.3 (NEIS)
4.	eSn	A	18 03 24	Central Italy 42.72 N 12.85 E
				H = 17 59 45.1 h = 5 km
				D = 7.97 Az = 354 (ISC)
4.	eP	A	20 26 14	Fox Islands, Aleutian Isl.
				51.90 N 170.86 W
				H = 20 14 19.4 h = 39 km MB = 4.5
				D = 77.81 Az = 358.4 (NEIS)
5.	eP	A	00 24 41	Ryukyu Islands 26.71 N 128.82 E
	e	A	25 07	H = 00 11 59.6 h = 20 km MB = 4.8
	LmH	B	01 07.4	D = 85.16 Az = 325.4 (NEIS)
	LmV	B	07.9	LmH B 16s 1.1 μm M = 5.4
				LmV B 13 1.2 μm 5.5

Day	Phase		h m s	Remarks
5.	ePn	A	05 24 04	Central Italy 43.06 N 12.63 E
	ePg	A	24 38	H = 05 22 10.4 h = normal MB = 4.2
	eSn	A	25 34	D = 7.62 Az = 355.1 (NEIS)
	eSg	A	26 14	
5.	eP	AB	12 10 29.5	Peru - Brazil Border Region
	epP	B	11 04	7.69 S 74.45 W
	ePP	C	14 30	H = 11 57 31.3 h = 162 km MB = 6.0
	epPP	C	14 50	D = 93.39 Az = 39.5 (NEIS)
	iSKS	B	20 52	PV A 1.8s 304.0nm M = 6.2
	eS	B	21 26	PV B 8 2.8/um 6.5
	eSP	C	22 30	SKSH B 12 7.5/um
	ePKKP	A	27 38	PKKPV A 1.2 28.5nm
	eSS	B	27 54	LmH B 17 2.1/um
	esSS	C	28 52	LmV B 17.5 3.8/um
	LmH	B	49.8	
	LmV	B	49.8	
5.	eP	A	23 10 31	Kurile Islands 46.07 N 149.00 E
				H = 22 58 56.3 h = 179 km MB = 4.8
				D = 76.86 Az = 333.7 (NEIS)
				PV A traces
6.	eP	A	14 11 18	Panama - Costa Rica Border Region
	LmH	B	51.8	8.30 N 82.85 W
	LmV	B	51.8	H = 13 58 38.6 h = 46 km MB=5.4 MS=5.4
				D = 86.47 Az = 39.5 (NEIS)
				LmH B 18s 1.7/um M = 5.5
				LmV B 18 1.8/um 5.5
6.	eP	A	17 22 27	Mariana Islands Region 21.61 N 145.67 E
	ePP	A	26 24.5	H = 17 08 53.8 h = normal MB = 5.4
				D = 97.45 Az = 332.5 (NEIS)
				PV A traces
6.	ePn	A	17 50 59	Central Italy 42.80 N 12.90 E
	eSn	A	51 30	H = 17 49 03.7 h = normal MB = 4.0
	eSg	A	52 26	D = 7.90 Az = 354.1 (NEIS)

Day	Phase		h m s	Remarks
7.	ePKP	A	05 49 20	Tonga Islands 15.10 S 173.48 W H = 05 29 46.0 h = normal MB=4.8 MS=4.7 D = 144.31 Az = 354.4 (NEIS)
7.	eiP	AB	07 46 06.5	Fox Islands, Aleutian Is.
	eS	B	56 00	51.86 N 170.80 W
	LmH	B	08 30.8	H = 07 34 11.0 h = normal MB=5.5 MS=5.8
	LmV	B	30.9	D = 77.85 Az = 358.4 (NEIS) PV A 1.3s 87.4nm M = 5.6 LmH B 17 4.2/um 5.8 LmV B 16 4.0/um 5.9
7.	eP	A	14 05 49.5	Burma - India Border Region 23.93 N 93.88 E H = 13 54 58.0 h = normal MB = 4.8 D = 67.08 Az = 316.8 (NEIS) PV A 1.1s 10.1nm M = 4.8
7.	ePP	A	14 47 36	Solomon Islands 5.65 S 154.34 E H = 14 26 52.3 h = 93 km MB = 5.4 (NEIS) D = 125.42
7.	eP	A	22 14 32	Near Islands, Aleutian Is.
	e	A	14 41	51.68 N 174.76 E H = 22 02 40.2 h = 33 km MB=5.0 MS=4.9 D = 77.06 Az = 349.1 (NEIS) PV A 1.0s 19.7nm M = 5.1
8.	eP	A	00 31 52	Iceland 63.99 N 22.82 W H = 00 26 53.6 h = normal MB = 4.3 D = 22.54 Az = 110.0 (NEIS) PV A 1.5s 20.1nm M = 4.4
8.	eP	A	01 05 37	Iceland Region 63.7 N 23.1 W H = 01 00 36 h = 0 km D = 22.56 Az = 109 (ISC) PV A 1.5s 15.1nm

Day	Phase		h m s	Remarks
8.	eP	A	01 31 34	Iceland Region 63.75 N 22.57 W
	LmV	B	42.9	H = 01 26 34.5 h = 28 km MB = 4.6
				D = 22.35 Az = 109.8 (NEIS)
				PV A 1.0s 15.8nm M = 4.4
				LmV B 13 0.7/um 4.4
8.	eP	A	01 51 26	Iceland 64.00 N 22.78 W
				H = 01 46 28.7 h = normal MB = 4.4
				D = 22.52 Az = 110.1 (NEIS)
8.	eP	A	03 55 34	Iceland 64.01 N 22.75 W
				H = 03 50 35.4 h = 15 km MB = 4.1
				D = 22.51 Az = 110.2 (NEIS)
8.	eP	A	04 01 12	Iceland Region 63.93 N 22.06 W
				H = 03 56 13.9 h = 34 km MB = 4.3
				D = 22.20 Az = 110.9 (NEIS)
				PV A 1.5s 20.1nm M = 4.3
8.	eP	A	06 52 56	Iceland Region 63.79 N 22.52 W
				H = 06 47 54.4 h = 11 km MB = 4.3
				D = 22.34 Az = 109.9 (NEIS)
				PV A 1.3s 15.3nm M = 4.3
8.	eP	A	07 10 14.5	Iceland 64.01 N 22.51 W
				H = 07 05 16.8 h = normal MB = 4.3
				D = 22.44 Az = 110.7 (NEIS)
				PV A 1.4s 14.0nm M = 4.2
8.	e	A	18 07 45	Tonga Islands Region 18.10 S 172.73 W
				H = 17 47 56.2 h = normal MB = 4.9
				D = 147.34 Az = 354.9 (NEIS)
9.	eP	A	04 37 24.5	Northern Yukon Territory, Canada
				65.16 N 134.10 W
				H = 04 27 08.9 h = 34 km MB = 4.3
				D = 61.53 Az = 24.1 (NEIS)

Day	Phase		h m s	Remarks
9.	ePKHKP	A	10 06 18	Tonga Islands 20.25 S 174.28 W
	ePKP2	A	06 23	H = 09 46 32.1 h = normal MB=5.4 MS=4.9
				D = 149.31 Az = 352.6 (NEIS)
				PKP2V A 2.2s 120.0nm
9.	eP	A	11 02 32	Mascarene Islands Region 17.02 S 66.87 E
				H = 10 50 07.9 h = normal MB = 5.0
				D = 82.94 Az = 328.2 (NEIS)
9.	ePn	A	12 15 22	Austria 48.17 N 16.86 E
	ePg	A	15 37	H = 12 14 15.8 h = 10 km
	eSg	A	16 29	D = 4.23 Az = 307.9 (NEIS)
9.	ePn	A	13 02 39	Northern Italy 46.78 N 12.82 E
	eSn	A	03 29	H = 13 01 40.4 h = normal
	eSg	A	03 51	D = 3.95 Az = 348.8 (NEIS)
10.	ePKHKP	A	01 23 16.5	Fiji Islands Region 19.95 S 178.54 W
	ePKP2	A	23 21.5	H = 01 04 37.9 h = 626 km MB = 4.7
				D = 148.33 Az = 347.7 (NEIS)
				PKHKPV A 1.0s 19.7nm
10.	ePn	A	01 36 38	D = 2.4
	eSg	A	37 16	
10.	eP	A	01 48 53.5	Hindu Kush Region 36.46 N 70.48 E
	epP	A	49 39.5	H = 01 41 06.0 h = 204 km MB = 5.5
	e	C	51 48	D = 43.81 Az = 308.1 (NEIS)
	eS	C	55 08	PV A 1.4s 158.1nm M = 5.3
	esS	C	56 24	
	eSS	C	58 32	
10.	eP1	A	02 39 50	North Atlantic Ridge 30.43 N 41.94 W
	eP2	A	39 55	H = 02 31 39.8 h = normal MB=5.0 MS=4.9
	LmH	C	55.2	D = 44.33 Az = 47.1 (NEIS)
	LmV	C	55.2	P2V A 2.0s 51.3nm M = 5.0
				LmH C 20 1.8 μm 5.0
				LmV C 18 2.5 μm 5.3

Day	Phase		h m s	Remarks
10.	e	A	11 37 23	
10.	eP	A	13 14 40	Off Coast of Hokkaido, Japan 42.50 N 146.54 E H = 13 02 36.3 h = 36 km MB = 5.0 D = 79.25 Az = 332.7 (NEIS)
10.	eP	A	19 58 37	West Pakistan 27.58 N 65.20 E H = 19 50 12.6 h = 43 km MB = 4.5 D = 46.32 Az = 314.9 (NEIS)
11.	eP	A	13 35 36	Off East Coast of Kamchatka 55.38 N 163.01 E H = 13 24 16.2 h = normal MB=4.7 MS=4.3 D = 71.69 Az = 341.3 (NEIS)
11.	ePKP epPKP	A A	17 30 21 30 48	Tonga Islands 17.48 S 174.10 W H = 17 10 55.4 h = 158 km MB = 4.7 D = 146.60 Az = 353.4 (NEIS) h = 95 km PKPV A traces
12.	ePn eSg	A A	01 36 43 38 10	Yugoslavia 45.8 N 15.1 E H = 01 35 23 h = 0 km D = 5.39 Az = 336 (ISC) PnV A 0.7s 13.4nm
13.	e(PP) e	A A	07 48 08 48 30	Philippine Islands Region 4.91 N 127.37 E H = 07 30 02.3 h = 90 km MB = 5.4 (NEIS) D = 102.13 PPV A 1.7s 24.2nm M = 5.6
13.	ePn ePg e eSn eSg	A A A A A	22 51 05.5 51 28.5 51 36 52 00 52 28	Northern Italy 45.74 N 12.39 E H = 22 49 52.3 h = normal D = 4.94 Az = 354.3 (NEIS)

Day	Phase		h m s	Remarks
13.	ePg	A	23 14 19	Probably Northern Italy
	e	A	14 27	D c. 4.9
	eSn	A	14 52	
	eSg	A	15 19	
14.	eP	A	02 39 55	Greece 38.28 N 20.75 E
	LmV	B	46.4	H = 02 36 38.4 h = normal MB = 5.3
	LmH	B	46.5	D = 13.97 Az = 335.2 (NEIS)
				PV A 0.9s 31.1nm M = 5.1
				LmH B 13 4.2 µm 4.7
				LmV B 12 2.9 µm
14.	e	A	06 34 54.5	
14.	eP	A	12 43 24.5	Kurile Islands 47.27 N 154.05 E
				H = 12 31 31.8 h = normal MB = 5.0
				D = 77.24 Az = 336.5 (NEIS)
				PV A 1.4s 42.8nm M = 5.3
14.	eP	A	20 20 52	Halmahera 2.62 N 128.41 E
	ePP	A	25 21	H = 20 07 10.0 h = 218 km MB = 5.5 (NEIS)
				D = 104.57
14.	eiP	A	21 32 35	Greece 38.61 N 20.41 E
				H = 21 29 19.8 h = 11 km MB = 5.0
				D = 13.56 Az = 335.5 (NEIS)
				PV A 0.9s 33.1nm M = 5.5
15.	ePKHKP	A	17 07 17	Tonga Islands 17.97 S 174.22 W
				H = 16 47 35.7 h = normal MB = 5.1 MS=4.9
				D = 147.07 Az = 353.2 (NEIS)
				PKHKPV A 1.6s 35.7nm
15.	eP	A	18 29 19	Norwegian Sea 74.60 N 12.72 E
				H = 18 24 07.6 h = normal MB = 4.6
				D = 24.05 Az = 181.7 (NEIS)

Day	Phase		h m s	Remarks
15.	e	A	20 59 05.5	Central Italy 22.43N 11.7 E
	e	A	59 08	H = 20 57 01.0 h = 0 km
	e	A	59 48	D = 7.10 Az = 359 (ISC)
	eSn	A	21 00 05	
16.	ePKIKP	A	05 30 40	Solomon Islands 7.48 S 155.97 E
				H = 05 11 49.3 h = 150 km MB = 4.8
				D = 127.79 Az = 332.0 (NEIS)
16.	eP	A	06 30 46.5	Eastern Kazakh SSR 49.76 N 78.06 E
	ePn	A	32 19	H = 06 23 02.4 h = normal MB = 5.0
				D = 41.24 Az = 297.7 (NEIS)
				Underground explosion (UPP)
				PV A 0.8s 23.1nm M = 5.0
16.	eP	A	06 48 46	Eastern Kazakh SSR 49.82 N 78.12 E
				H = 06 41 02.4 h = normal MB = 4.8
				D = 41.24 Az = 297.7 (NEIS)
				Underground explosion (UPP)
				PV A 0.7s 23.0nm M = 5.0
16.	ePP	C	08 15 10	Easter Island Region 24.88 S 112.12 W
	eSKP	C	16 32	H = 07 53 56.8 h = 33 km MB = 5.1 (NEIS)
	eSS	C	32 40	D = 130.07
	eSSS	C	37 25	LmH C 16s 1.2 µm M = 5.7
	LmH	C	09 13.0	LmV C 16 2.5 µm 6.0
	LmV	C	17.4	
16.	ePn	A	17 20 20	Federal Republic of Germany
	eSn	A	20 57	49.72 N 6.41 E
	eSg	A	21 13	H = 17 19 23.2 h = normal
				D = 3.48 Az = 72.5 (NEIS)
16.	ePn	A	17 54 08	Federal Republic of Germany
	e	A	54 13	49.75 N 6.45 E
	e	A	54 59	H = 17 53 14.3 h = normal
	e	A	55 04	D = 3.44 Az = 73.0 (NEIS)
				PnV A traces

Day	Phase		h m s	Remarks
16.	ePKIKP	A	23 19 05	Fiji Islands Region 19.53 S 178.12 W
	ePKHKP	A	19 09	H = 23 00 30.5 h = 624 km MB = 4.8
				D = 148.01 Az = 348.3 (NEIS)
				traces
17.	ePKHKP	A	15 55 11	West of Macquarie Island
	ePKP2	A	55 18	54.14 S 143.43 E
	LmH	C	17 03.0	H = 15 35 21.2 h = normal MB=5.1 MS=5.8
	LmV	C	03.3	D = 150.84 Az = 283.1 (NEIS)
				LmH C 22s 2.9 µm M = 6.0
				LmV C 22 1.8 µm 5.8
17.	ePKP	A	21 59 04	Tonga Islands 20.94 S 175.22 W
	e	A	59 16	H = 21 39 11.1 h = normal MB = 4.8
				D = 149.86 Az = 351.3 (NEIS)
				PKPV A 1.6s 27.4nm
17.	ePKHKP	A	23 21 40	Tonga Islands 20.53 S 175.34 W
	ePKP2	A	21 45	H = 23 01 52.6 h = 31 km MB = 5.4 MS=5.2
				D = 149.44 Az = 351.3 (NEIS)
				PKP2V A 1.6s 93.5nm
18.	eP	A	08 04 16	Mongolia 48.40 N 103.11 E
	LmV	B	30.0	H = 07 54 40.4 h = normal MB=5.0 MS=5.1
	LmH	B	30.5	D = 55.68 Az = 309.6 (NEIS)
				PV A 1.8s 74.4nm M = 5.4
				LmH B 13.5 2.6 µm 5.5
				LmV B 14 3.0 µm 5.6
18.	eP	A	20 16 18	Norwegian Sea 67.85 N 10.35 E
				H = 20 12 15.1 h = normal MB = 4.6
				D = 17.27 Az = 177.3 (NEIS)
19.	ePKP	A	11 17 30	Fiji Islands 17.47 S 179.04 E
				H = 10 59 01.9 h = 646 km MB = 4.5
				D = 145.40 Az = 345.9 (NEIS)
				PKPV A 1.8s 54.0nm

Day	Phase		h m s	Remarks
19.	ePKIKP	A	15 52 23.5	Santa Cruz Islands 12.31 S 166.67 E
				H = 15 33 14.5 h = 133 km MB = 4.9
				D = 136.67 Az = 337.0 (NEIS)
				PKIKPV A 1.0s 9.9nm
19.	eP	AB	16 13 23.5	Panama 7.38 N 78.65 W
	eS	C	24 05	H = 16 00 49.0 h = 13 km MB=5.4 MS=5.8
	ePS	C	24 44	D = 84.52 Az = 39.8 (NEIS)
	eSS	C	29 28	PV A 1.5s 40.2nm M = 5.4
	LmH	B	45.0	LmH B 21 6.1 /um 6.0
	LmV	B	45.0	LmV B 22 4.7 /um 5.9
20.	-eiPKP	A	02 59 23.5	Fiji Islands Region 15.39 S 177.09 W
				H = 02 40 31.3 h = 374 km MB = 5.1
				D = 144.14 Az = 350.5 (NEIS)
				PKPV A 1.3s 78.6nm
20.	ePKP	A	14 49 27	Tonga Islands 20.15 S 173.96 W
				H = 14 29 47.6 h = 117 km MB = 4.8
				D = 149.25 Az = 353.1 (NEIS)
				PKPV A 1.8s 40.6nm
20.	eP	A	15 12 29.5	Greece-Albania Border Region
	LmH	C	17.8	39.74 N 20.38 E
	LmV	C	18.4	H = 15 09 29.0 h = 12 km MB = 4.8
				D = 12.53 Az = 333.4 (NEIS)
				PV A 1.3s 21.8nm M = 5.2
				LmH C 14 2.4 /um 4.4
				LmV C 14 3.1 /um
20.	eiP	A	16 49 46.5	Northwest of Kurile Islands
	epP	A	51 22.5	49.83 N 149.68 E
				H = 16 38 55.6 h = 416 km MB = 5.0
				D = 73.69 Az = 333.7 (NEIS)
				h = 436 km
				PV A 1.4s 69.7nm M = 5.1

Day	Phase		h m s	Remarks
21.	eP	A	00 39 29	Panama 7.34 N 78.64 W H = 00 26 53.8 h = normal MB = 5.1 D = 84.55 Az = 39.8 (NEIS) PV A 1.3s 19.6nm M = 5.1
21.	iPg	A	01 47 50.3	Vogtland, German Democratic Republic
	iSg	A	49 58.3	50.5 N 12.6 E H = 01 47 39 h = 0 km D = 0.62 Az = 282 (ISC)
21.	ePKP	A	08 48 33	Samoa Islands Region 14.56 S 175.23 W
	LmH	B	09 54.5	H = 08 28 55.9 h = normal MB=5.6 MS=6.1
	LmV	B	10 03.4	D = 143.59 Az = 352.7 (NEIS) PKPV A 1.8s 67.5nm LmH B 20 4.4 /um M = 6.2 LmV B 16 3.1 /um 6.2
21.	e(Pn)	A	15 16 24.5	
	e	A	16 33	
	e	A	16 39.5	
	eSn	A	17 03.5	
	e(Sg)	A	17 31	
21.	eP	A	20 51 02	Burma-China Border Region 25.84 N 97.78 E H = 20 40 01.7 h = normal MB = 4.9 D = 68.12 Az = 316.8 (NEIS) traces
22.	ePKP2	A	02 38 55	South of Fiji Islands 26.89 S 176.18 W H = 02 18 36.9 h = normal MB = 5.1 D = 155.54 Az = 348.0 (NEIS)
22.	eP	A	04 57 38	North Eastern China 41.7 N 123.30 E H = 04 46 18.9 h = 0 km MB = 5.0 D = 70.43 Az = 321 (ISC)

Day	Phase		h m s	Remarks
22.	+iPKP	A	17 02 44	Fiji Islands Region 17.62 S 179.02 W H = 16 44 05.3 h = 559 km MB = 5.1 D = 145.97 Az = 347.9 (NEIS) PKPV A 2.0s 184.0nm
22.	ePKP2	A	18 26 54	Kermadec Islands Region 26.95 S 176.30 W H = 18 06 36.2 h = normal MB=5.2 MS=5.0 D = 155.58 Az = 347.8 (NEIS)
23.	ePKP LmV	A C	01 23 37.5 02 23.8	Samoa Islands Region 14.59 S 175.67 W H = 01 04 02.7 h = normal MB=5.4 MS=5.6 D = 143.56 Az = 352.2 (NEIS) LmV C 22s 1.0 µm M = 5.5
23.	eP1 eP2 eP3 LmH LmV	A A A B B	05 27 34 27 39.5 27 43.5 39.5 40.5	Eastern Caucasus 43.13 N 47.05 E H = 05 22 08.4 h = normal MB=4.9 MS=5.0 D = 25.17 Az = 299.8 (NEIS) P1V A 0.8s 15.4nm M = 4.6 P2V A 1.0 19.7nm 4.6 P3V A 1.6 60.4nm 4.9 LmH B 14 1.9 µm 4.8 LmV B 11 2.3 µm 5.1
23.	+eP	A	09 55 14	Nepal 29.41 N 81.39 E H = 09 45 42.8 h = 45 km MB = 5.2 D = 55.33 Az = 313.4 (NEIS) PV A 1.1s 32.3nm M = 5.3
23.	+ePKP	A	11 35 38.5	Fiji Islands Region 16.21 S 176.66 W H = 11 16 48.3 h = 421 km MB = 4.6 D = 145.01 Az = 350.8 (NEIS) PKPV A 0.9s 23.4nm
23.	e(P)	A	16 35 46	Crete 35.39 N 26.08 E H = 16 31 24.5 h = normal MB = 4.5 D = 18.50 Az = 329.9 (NEIS)

Day	Phase		h m s	Remarks
23.	e(P)	A	16 35 46	Crete 35.39 N 26.08 E
				H = 16 31 24.5 h = normal MB = 4.5
				D = 18.50 Az = 329.9 (NEIS)
23.	eP	A	23 55 06.5	South of Panama 5.26 N 82.54 W
	e	A	55 41.5	H = 23 42 14.5 h = normal MB=5.1 MS=5.0
	LmV	C	24 27.3	D = 88.61 Az = 39.4 (NEIS)
				LmV C 22s 1.6 /um M = 5.4
24.	eP	AC	02 22 51.5	Guatemala 14.26 N 90.08 W
	epP	A	23 28	H = 02 10 25.4 h = 155 km MB = 5.4
	LmV	C	54.0	D = 86.35 Az = 38.7 (NEIS)
				h = 146 km
				pPV A 1.6s 159.4nm
24.	eP1	A	07 08 45	Southern Sumatra 2.31 S 99.04 E
	eP2	A	08 48	H = 06 55 47.1 h = normal MB=5.8 MS=6.8
	eS	B	19 36	D = 90.13 Az = 320.5 (NEIS)
	LmH	B	08 05.3	P1V A 2.4s 166.5nm M = 5.9
	LmV	B	12.4	P2V A 1.9 310.6nm 6.3
				LmH B 16 12.4 /um 6.4
				LmV B 15 11.4 /um 6.4
24.	eP	A	10 32 02	Turkey 37.52 N 29.80 E
				H = 10 27 42.8 h = 25 km MB = 4.6
				D = 18.45 Az = 321.1 (NEIS)
24.	eP	A	20 03 16	Hindu Kush Region 36.40 N 70.39 E
				H = 19 55 29.8 h = 206 km MB = 4.6
				D = 43.79 Az = 308.1 (NEIS)
25.	ePKIKP	A	02 36 21	New Hebrides Islands 14.30 S 167.25 E
				H = 02 17 14.4 h = 179 km MB = 5.0
				D = 138.72 Az = 336.5 (NEIS)
25.	+eiP	A	03 01 03.5	Near Islands, Aleutian Is.
	eS	C	10 52	51.70 N 174.64 E
	LmH	B	44.6	H = 02 49 13.0 h = 40 km MB=5.7 MS=5.8

Day	Phase		h m s	Remarks
cont. 25.	LmV	B	03 47.3	D = 77.02 Az = 349.0 (NEIS) PV A 1.4s 134.9nm M = 5.8 LmH B 16 3.9 μm 5.8 LmV B 16 3.6 μm 5.8
25.	eP	A	08 06 37	Near Islands, Aleutian Is. 51.73 N 174.49 E H = 07 54 46.0 h = 37 km MB=5.1 MS=4.8 D = 76.97 Az = 348.9 (NEIS) PV A 1.3s 21.8nm M = 5.0
25.	eP	A	09 30 14	South Atlantic Ridge 28.95 S 13.13 W H = 09 17 48.3 h = 0 km MB = 4.8 D = 82.22 Az = 16 (ISC)
25.	eP	A	11 26 25	Unimak Island Region 54.08 N 163.38 W H = 11 14 40.7 h = 30 km MB = 4.4 D = 75.57 Az = 3.3 (NEIS)
25.	ePKIKP ePKP2 e(PP)	A A A	12 52 28 53 07 57 03.5	Kermadec Islands Region 31.90 S 179.84 E H = 12 33 09.6 h = 327 km MB = 5.1 D = 159.34 Az = 338.4 (NEIS) PKP2V A 1.4s 32.5nm PPV A traces
25.	eP epP	A A	15 13 48 15 38	Northwest of Kurile Islands 52.51 N 152.05 E H = 15 03 16.7 h = 524 km MB = 4.8 D = 71.92 Az = 334.7 (NEIS) h = 543 km
26.	eP e	A A	14 05 59.5 06 07	Crete 35.00 N 23.22 E H = 14 01 49.2 h = normal MB = 4.4 D = 17.77 Az = 335.2 (NEIS) PV A traces
26.	eP	A	18 43 33	Southern Iran 29.50 N 52.73 E H = 18 36 21.9 h = normal MB = 4.8 D = 37.26 Az = 316.3 (NEIS)

Day	Phase		h m s	Remarks
27.	ePKP	A	01 11 56	Fiji Islands Region 16.41 S 176.20 W
				H = 00 53 00.7 h = 380 km MB = 4.6
				D = 145.27 Az = 351.3 (NEIS)
				PKPV A 1.2s 12.2nm
27.	+iP	A	05 54 48.8	Eastern Kazakh SSR 49.96 N 79.05 E
	ePr	A	56 26	H = 05 46 56.8 h = 0 km MB=5.6 MS=4.7
				D = 41.71 Az = 297.9 (NEIS)
				Underground explosion MB = 6.4 (UPP)
				PV A 1.0s 78.9nm M = 5.4
				PnV A 1.1 32.3nm
27.	ePKIKP	A	11 48 47	Kermadec Islands 29.29 S 177.45 W
	ePKP2	A	49 20	H = 11 28 57.8 h = 61 km MB = 5.0 (NEIS)
				D= 157.60
				PKP2V A 1.5s 30.2nm
27.	ePKP2	A	12 16 45	Kermadec Islands 29.40 S 177.34 W
				H = 11 56 18.1 h = 22 km MB = 4.9 (NEIS)
				D = 157.73
28.	eP	A	00 23 35	Alaska Peninsula 54.19 N 162.77 W
				H = 00 11 53.7 h = 44 km MB = 4.6
				D = 75.43 Az = 3.7 (NEIS)
				PV A 1.1s 20.2nm M = 5.0
28.	+eiP	AB	12 20 08.5	West Pakistan 35.05 N 72.87 E
	ePP	C	22 00	H = 12 11 43.8 h = 22 km MB=6.0 MS=6.2
	ePPP	C	22 25	D = 46.20 Az = 309.3 (NEIS)
	eS	C	26 45	PV A 2.2s 577.0nm M = 6.1
	eSS	C	30 16	PV B 8 4.6 µm 6.5
	LmH	B	40.3	LmH B 17 17.4 µm 6.1
	LmV	B	43.5	LmV B 15 22.2 µm 6.3
28.	eP	A	24 02 14	Mexico-Guatemala Border Region
	epP	A	02 59.5	14.97 N 91.06 W
				H = 23 49 51.0 h = 186 km MB = 4.6
				D = 86.39 Az = 38.5 (NEIS)
				h = 185 km

Day	Phase		h m s	Remarks
29.	eP1	AB	03 54 44.5	Iceland 64.51 N 17.63 W
	eP2	AB	54 47.5	H = 03 50 06.1 h = normal MB = 5.2
	eS	B	58 40	D = 20.67 Az = 118.2 (NEIS)
	LmV	B	04 03.8	P1V A 1.7s 90.9nm M = 4.9
	LmH	B	04.8	P2V A 1.6 258.0nm 5.3
				LmH B 14 4.5 µm 5.0
				LmV B 20 4.9 µm 5.0
29.	eP	A	05 07 04	Tadzhik SSR 39.20 N 71.81 E
				H = 04 59 09.6 h = 69 km MB = 5.0
				D = 43.00 Az = 305.9 (NEIS)
29.	-iP	AB	18 35 49.0	Southern Alaska 61.60 N 150.51 W
	epP	AB	36 08	H = 18 25 00.7 h = 67 km MB = 5.6
				D = 67.20 Az = 12.2 (NEIS)
				h = 77 km
				PV A 1.1s 230.0nm M = 6.1
				pPV A 1.3 91.7nm
30.	eP	A	03 44 02	Central Alaska 61.98 N 149.69 W
	epP	A	44 18.5	H = 03 33 16.6 h = 62 km MB = 5.1
				D = 66.74 Az = 12.8 (NEIS)
				h = 64 km
				PV A 1.0s 21.6nm M = 5.1
				pPV A 0.9 40.9nm
30.	eP	A	04 55 39	Hindu Kush Region 35.99 N 69.72 E
	ePP	A	57 22	H = 04 47 44.3 h = 116 km MB = 5.3
				D = 43.61 Az = 308.4 (NEIS)
				PV A 1.4s 69.8nm M = 5.2
				PPV A 1.2 40.7nm 5.0
30.	eP	A	15 28 39.5	Szechwan Province, China
				28.85 N 103.95 E
				H = 15 17 29.2 h = normal MB = 4.7
				D = 69.67 Az = 317.3 (NEIS)

Day	Phase		h m s	Remarks
30.	eP	A	24 09 12	<u>Off East Coast of Honshu, Japan</u> 39.53 N 143.18 E H = 23 57 03.2 h = 51 km MB = 4.6 D = 80.67 Az = 331.1 (NEIS)
31.	eP	A	04 48 58.5	<u>Andreanof Islands, Aleutian Is.</u> 52.53 N 179.26 W H = 04 37 31.2 h = 228 km MB = 5.1 D = 76.79 Az = 352.9 (NEIS)
31.	eP	A	20 28 14.5	<u>Guatemala</u> 14.15 N 91.92 W H = 20 15 32.8 h = 75 km MB = 5.4 D = 87.55 Az = 38.3 (NEIS) PV A 1.3s 21.8nm M = 5.1
31.	eP	AB	20 33 54	<u>Guatemala</u> 14.13 N 91.82 W
	eSKS	B	44 26	H = 20 21 09.1 h = 39 km MB=5.7 MS=6.1
	eSS	B	50 24	D = 87.51 Az = 38.3 (NEIS)
	eSSS	B	54 32	PV A 1.0s 19.7nm M = 5.3
	LmH	B	21 16.0	PV B 14 1.5/um 6.1
	LmV	B	16.0	LmH B 18.5 6.8/um 6.2
				LmV B 19 6.8/um 6.1
31.	e	A	21 02 44	<u>Czechoslovakia</u> 49.9 N 18.7 E H = 21 00 25 h = 0 km (ISC) D = 4.6

Bei Fragen zur Produktsicherheit wenden Sie sich bitte an:
If you have any questions regarding product safety,
please contact:

Walter de Gruyter GmbH
Genthiner Straße 13
10785 Berlin
productsafety@degruyterbrill.com